Springer Proceedings in Business and Economics

Springer Proceedings in Business and Economics brings the most current research presented at conferences and workshops to a global readership. The series features volumes (in electronic and print formats) of selected contributions from conferences in all areas of economics, business, management, and finance. In addition to an overall evaluation by the publisher of the topical interest, scientific quality, and timeliness of each volume, each contribution is refereed to standards comparable to those of leading journals, resulting in authoritative contributions to the respective fields. Springer's production and distribution infrastructure ensures rapid publication and wide circulation of the latest developments in the most compelling and promising areas of research today.

The editorial development of volumes may be managed using Springer's innovative Online Conference Service (OCS), a proven online manuscript management and review system. This system is designed to ensure an efficient timeline for your publication, making Springer Proceedings in Business and Economics the premier series to publish your workshop or conference volume.

More information about this series at http://www.springer.com/series/11960

Michal Suchanek
Editor

Transport Development Challenges in the 21st Century

Proceedings of the 2019 TranSopot Conference

Springer

Editor
Michal Suchanek
Faculty of Economics
University of Gdańsk
Sopot, Poland

ISSN 2198-7246 ISSN 2198-7254 (electronic)
Springer Proceedings in Business and Economics
ISBN 978-3-030-50009-2 ISBN 978-3-030-50010-8 (eBook)
https://doi.org/10.1007/978-3-030-50010-8

This Springer imprint is published by the registered company Springer Nature Switzerland AG
The registered company address is: Gewerbestrasse 11, 6330 Cham, Switzerland

Contents

Logistics Startups and Logistics Service Providers: Competitors or Partners in Exploration?

Marzenna Cichosz

Abstract The logistics industry is facing digital disruption along its entire value chain. Agile, innovative startups develop an expanding range of technology-driven solutions for industry pain points. The purpose of this chapter is to look at logistics startups, their characteristics and taxonomy, but foremost their relationships with incumbent logistics service providers (LSPs). Within this chapter the author analyzes whether startups and incumbent LSPs are competitors or partners in exploration and how LSPs should cooperate with startups to fuel their innovativeness and growth. The chapter is based on the qualitative analysis of logistics startups websites as well as semi-structured interviews with three leading LSPs. At the theoretical level, the research contributes to enriching logistics studies by presenting logistics startups and analyzing their impact on the logistics service industry. At a managerial level, the findings provide illustration on how to collaborate with startups to increase LSPs' innovativeness.

Keywords Startup · Technology · Digitalization · Open innovation · Venture capital · Intermediation platforms · Drones · LSP

1 Introduction

Having transformed consumer-oriented industries such as retail, entertainment, and travel, digital startups are beginning to make inroads in commercial-oriented industries, including the logistics service industry. According to the Boston Consulting Group, from 2012 through 2017 venture capitals funds invested more than 3.3 billion USD in digital shipping and logistics startups [1]. Oliver Wyman [2, p. 2] reports that "around every fifth day, a new logistics startup is founded" and warns

M. Cichosz (✉)
Collegium of Management and Finance, SGH Warsaw School of Economics, Warsaw, Poland
e-mail: Marzenna.Cichosz@sgh.waw.pl

M. Suchanek (ed.), *Transport Development Challenges in the 21st Century*, Springer Proceedings in Business and Economics,
https://doi.org/10.1007/978-3-030-50010-8_1

that "these agile, innovative startups are disrupting logistics service industry along its entire value chain from freight forwarding, brokerage, and long-distance transportation, to warehousing, contract logistics, and last-mile delivery." They are capitalizing on the high number of transactions and large amounts of data being handled and generated by logistics players to develop an expanding range of technology-driven solutions to industry pain points.

Logistics incumbents, surrounded by logistics startups, feel the change in transportation and logistics. Various players from various industries (e.g., technology companies operating in the retail industry such as Amazon, vehicle manufacturers such as Daimler, or electronic platform operators offering logistics services in the crowd logistics model such as Deliv) have entered the race for tomorrow's leading market position in the logistics service industry [3]. Therefore, (re)action from established logistics service providers (LSPs) is strongly required. They have to overcome barriers and increase the level of their innovativeness, especially in the technological area to stay competitive. Therefore, the following research questions arise:

- Are logistics startups threatening competitors of LSPs?
- Should LSPs use the logistics startups as a springboard to fuel their innovativeness?
- How could LSPs collaborate with the startups to be most effective in exploration?

Despite the significance of this topic for the future of the logistics service industry, up to now digitalization and logistics startups have barely gained any attention in logistics literature. So far, these topics have mainly been studied by consulting companies and research departments of the biggest LSPs. The purpose of this chapter is to investigate logistics startups, their characteristics and taxonomy, but foremost their relationships with incumbent logistics service providers (LSPs) while exploring innovative solutions to the logistics industry's pain points. The chapter is based on the qualitative analysis of logistics startups websites as well as semi-structured interviews with three LSPs' managers. The deduction method is applied.

2 Theoretical Background

2.1 Logistics Startups

The term "startup" is understood in various ways. On the one hand, researchers identify startups as beginner companies that enter the market with limited resources and little business experience. On the other hand, startups have been perceived as highly innovative, ambitious, dynamic technological undertakings [4, 5].

The most popular definition of a startup was developed by Blank [6, p. 65]. He describes a startup as "a temporary organization formed to search for a repeatable and scalable business model." Blank claims that startups can be identified by their:

(1) ambitious goals towards becoming a large company, (2) continuous testing of business hypotheses, their verification, and possible modification of subsequent versions of the business model, and (3) a financing structure that includes funds obtained from external investors. Blank's definition does not contain any technological element. However, Skala [5, p. 16] notices that technological aspects appear implicitly as "(...) the scalability of the business model can be achieved mainly thanks to the automation of important operations (tasks)." Ries [7] describing startups emphasizes "the extreme uncertainty" in which they operate. Christensen and Raynor [8] see startups as organizations that create breakthrough innovations that in the long run can change market paradigms. Damodaran [9] points out a high potential for growth in the company's value in the future.

In Polish literature, Skala [5] presents that Glinka and Pasieczny [10] clearly indicate that a startup is a young company. Cieślik [11] points to startups as technology companies that operate in the ICT and Internet industries. Łuczak [12] considers the features related to the early stage of development, innovation, and growth potential, as well as the search for the optimal business model. Bursiak [13] suggests a 5-year time limit and private external funding. Finally, Latusek-Jurczak [14] focuses on rapid growth and scalability.

Concluding the review of startups' definitions, the list of main characteristics that distinguish startups from other enterprises is following:

- The young age of the enterprise and limited resources (especially financial)
- Innovation (startups offer innovative solutions in an innovative way)
- Development and scalability (startups are ambitious and fast-growing companies which are easy to scale up)
- Technology (startups are companies operating in the digital industry or, more broadly, technology companies)

Logistics startups are startups characterized by the same set of features; however, their focus is on logistics innovations at LSPs, i.e., radical as well as incremental ones if they are valuable for business operational efficiency or creating a better customer experience. As noticed by Kurpjuweit et al. [15, p. 2] "a logistics startup cannot be a newly founded company that simply replicates a conventional logistics business (e.g. starting a trucking business), but rather one that introduces an innovation for LSPs' themselves or to a subset of LSPs' customers." Kurpjuweit et al. [15, p. 2] define a logistics startup as "a new venture whose value creation process is closely linked to the logistical activities of LSPs." The clue of this approach is value creation by delivering logistics innovations.

2.2 *Logistics Innovation at LSPs*

Logistics innovation, which can occur within any service, process, or social system, is defined as "a new, helpful idea, procedure, or practice in logistics operation that is different from a company's current practice" [16, p. 361]. Logistics innovation can range from the basic to the complex, and can be applied to improve operational efficiency of the LSP or better serve customers [17].

Wagner [18] points out that an LSP's innovation activities embrace not only the development of a new logistics concept but also the adaptation and implementation of existing logistics concepts. As the logistics service industry stands at the low end of innovativeness, compared to other industries (e.g., [19–22]), LSPs can benefit from opening the innovation process to external sources of knowledge and innovation capabilities. Different studies in the area of logistics and supply chain management show that the external sources of innovation for LSPs could be:

- Customers (including final consumers in B2C logistics), e.g., [17, 20, 23, 24]
- Suppliers, e.g., [18, 25]
- Universities or research centers, e.g., [19], or even
- Competitors, e.g., [26]

The question is whether LSPs should use logistics startups to increase their innovativeness. And, furthermore, how they could cooperate with these young, small, fast-growing, technological enterprises to be most effective in the exploration process.

2.3 Large Corporations' Cooperation with Startups

Literature on cooperation and partnership between corporations and startups emphasizes that they are decidedly different organizations [27]. One has what the other lacks. Weiblen and Chesbrough [28, p. 66] observe: "The corporation has resources, scale, power, and the routines needed to run a proven business model efficiently. The startup has none of those, but typically has promising ideas, organizational agility, the willingness to take risks, and aspirations of rapid growth." The complementary nature of corporations and startups suggests that both can benefit from cooperating. In doing so, corporations receive support in exploration, as opposed to exploitation, while startups receive help to improve the execution of their nimble processes.

The analysis of how corporations can engage with startups shows that partnering with startups has matured over time, giving rise to programs initiated by corporations and third party specialists. The most popular form is a corporate accelerator (i.e., a company-supported program of limited duration that backs cohorts of startups during the new venture process via mentoring, education, and company-specific resources) [27]. According to Kohler, setting up a formalized corporate accelerator might result in:

- Corporation supports pilot project.
- Corporation becomes startup customer.
- Corporation becomes distribution partner.

- Corporation invests in startup.
- Corporation acquires startup.

While harnessing these cooperation opportunities can be beneficial for both sides, they are difficult to achieve because of the partners' asymmetry in power and structure. This is reflected in some of the phrases coined to describe these partnerships, such as "swimming with sharks" [29], "surviving bearhugs" [30], or "dancing with gorillas" [31]. Therefore, the latter (p. 9) suggest that "being seen as a 'safe' partner for a startup is likely to be of growing importance."

LSP's cooperation with startups was studied by Cichosz [3, 32] who refers mainly to large logistics players, clustered into: 3PLs, 4PLs, and CEP (Couriers Express Parcel) companies. Analyzing different types of LSPs' cooperation on open innovation, Cichosz [32] has applied a two-dimension model with partner selection and cooperation intensity dimensions to distinguish four types of cooperation, i.e., addressed or distributed solutions acquisition, and close or open partnerships.

3 Research Method

The chapter is based on qualitative studies, which, according to Edmondson and McManus [33], is a perfect methodological fit for the current stage of LSP-logistics startups cooperation concept development. The research adopts a two-stage effort.

In stage one, the author conducted a logistics startups' profiles analysis. In order to gain knowledge on the wide spectrum of logistics startups' solutions, the author visited CB Insights (www.cbinsights.com), one of the largest startup databases, and downloaded "The supply chain and logistics Tech Market Map." This map, transformed into Table 1, embraces 13 categories of supply chain and logistics startups. The author visited the websites of all startups listed in the table. Based on the website information and media available there, the author was able to understand the solution offered by the startup, its value for customers, and its business model. The information from CB Insights and the startups' websites supplemented with interviews conducted with three LSPs allowed for evaluating startups' competitive position in the logistics service market.

In stage two, the author conducted case study analysis, as it is a particularly suitable method for new topics [34]. For the analysis, the author selected three logistics companies which differ in their approach to logistics startups and are best described in professional industry publications. To gain personal insight, the author conducted semi-structured interviews on LSP-startup cooperation with managers from the case LSPs. The interviews took place in March 2019 and lasted on average 36 min each.

Table 1 The supply chain and logistics tech market players based on the CB Insight analysis

Categories	Startup examples
Digital freight shipping	Convoy, coLoadX, Flexport, Freight Hub, Fleet, Freightos, iContainer, Inttra, Nyshes, Shippabo, Xeneta
Sensors and assets tagging	Alien Technology, C3 AI & IoT Software, Omni-ID, Savi
Inventory management	Celect Ecomdash, Nextail, Optoro, TradeGecko
Blockchain	300cubits, ICIX, FoodLogiQ, ShipChain
Food supply chain	Clear Labs, Full Harvest, AgShift, BlueWrap, Apeel Sciences
Supply chain & logistics analytics	ClearMetal, FourKites, Shipamax, Project44, riskmethods Solutions
Trucking marketplace & fleet management	uShip, InstaFreight, RIVIGO, Everoad, Manbang Group, BlackBuck, Cargomatic, Transfix, KeepTruckin, FleetUp
Warehousing	6 River Systems, Clearpath Robotics, Fetch Robotics, Flexe, Log-Hub, CommonSense Robotics, Magazino, Exotec Solutions, Darkstore
ERP	Exostar, E2open, Nulogy, Sight Machine
E-commerce logistics	ShipBob, Shippo, Locus, Sendle, Narvar Concierge, Ecom Express, FineEx, FarEye, Happy Returns, Mandae, Bringg
Autonomous trucking	Embark, Kodiak Robotics, Starsky Robotics, TuSimple
Last mile logistics AV & drones	Matternet, Flirtey, Starship Technologies, Marble, Udely, Nuro
Last mile delivery services	Go-Jek, Postmates, Deliv, Lalamove, Ninjavan, Glovo

4 Findings and Discussion

The results of the study are structured around the research questions. The first section relates to logistics startups as threatening competitors of established LSPs. The second investigates options of LSPs-startups partnering on innovation and analyzes how these partnerships could be managed.

4.1 Logistics Startups Exploring Logistics Pain Points

The logistics service industry's main pain points are following [1]:

- High fragmentation of the transport and logistics (T&L) market
- High commoditization of service offerings
- Underutilization of logistics assets, i.e., trucks, warehouses
- Mostly manual processes and very low level of automation in logistics
- Low end-to-end visibility resulting from, e.g., lack of real-time cargo tracking
- Outdated customer service interfaces

In general, problems constitute therats or create opportunities for innovation. As most LSPs are not able to take advantage of these opportunities that creates a perfect situation for startups. The analysis shows that startups are the pith and marrow of innovativeness for the logistics service industry. Being lean (i.e., eliminating wasteful practices) and adopting agile development methods (i.e., experimenting, learning from failures, and iterating the whole R&D process), combined with the high willingness to take risks and grow fast, helps them to turn their promising ideas into innovation success, which thanks to technology, can be scaled up. This allows logistics startups to challenge the status quo and lay down the foundation for a new competitive landscape. To systemize logistics startups, Kurpjuweit et al. [15] divided the group into four main categories:

- Intermediation platform
- Software providers
- Hardware technologies
- CEP (Couriers Express Parcel) services

The results of the logistics startups' analysis and semi-structured interviews show that the biggest threat for incumbent LSPs are different types of intermediation platforms. These digital, cloud-based platforms are focused on matching supply and demand for logistics services (mainly in transport, e.g., Cargonexx, LoadFox, but also in storage and warehousing, e.g., Flexe, Log-Hub). With nearly zero physical assets and purely on the backbone of modern IT solutions, the platforms strive to take over the role of an efficient intermediary. Their main competitive advantage is creating full price transparency. This is a significant threat for different clusters of logistics players, i.e., carriers, forwarders, 3PLs, and 4PLs, who are afraid of price and margin erosion, which are already very low in the logistics service industry. Another threat for incumbent LSPs is the fact that platforms can change power distribution in the logistics service industry and increase freight carriers' independence from LSPs, as they can offer their free capacity directly to shippers. Furthermore, freight carriers can gain new customers, collect their feedback and offer additional value-added services such as track-and-trace, and in this way also cause LSPs' market share loss. The intermediation platforms are also threatening 4PLs, whose business model is based on the ability to connect partners' resources and offer a comprehensive services. Thanks to digital technology, intermediation platforms gain a competitive advantage over 4PL incumbents as they have access to a more extensive network of carriers and 3PLs and use advanced algorithms to calculate and predict rates, capacities, and means of optimization (e.g., Cargonexx).

Another example of an innovative platform threatening LSPs, mainly in CEP industry, is a crowdshipping platform such as PiggyBee and Deliv. Predominantly, these types of platforms target B2C customers that intend to ship small cargo units. A shipper procures transport services via a mobile or computer application directly from members of the crowd who provide those services as independent contractors using a personally owned vehicle. A crowdsourced fleet competes with traditional dedicated courier fleets in urban areas.

The situation with startups' competition looks differently in case of startups belonging to the category of software and hardware providers. The author considers them not threatening competitors who take away part of LSPs' market share, but rather innovators who aim to support T&L companies in improving their operational efficiency and enhancing customer service. Therefore, software providers offer solutions enabling value-added logistics services such as track-and-trace (e.g., Four Kites), fleet management and route optimization (e.g., Flutaro), blockchain (e.g., 300cubits), or big data analytics (e.g., riskmethods Solutions). Startups providing hardware have patents on tangible technology applicable in the logistics service industry which is often supported by software. The hardware technologies could be applied within the LSPs facilities like robots for handling warehouse picking process (e.g., 6 River Systems), or externally such as with fulfillment facilities (e.g., Darkstore), parcel lockers (e.g., ParcelLock) including mobile parcel lockers (e.g., Mobile Locker), autonomous vehicles for city deliveries (e.g., Nuro, Marble), or drones for delivering packages in hard-to-reach places (e.g., Matternet, Flirtey). Hardware technology providers are very active in CEP services.

4.2 LSPs' Cooperation with Startups

The growth and increasing viability of logistics startups forced large and established LSPs to join the competition for logistics innovations. In general, they take a two-fold approach. On the one hand, they establish their own intermediation platforms to avoid missing opportunities and promote their own assets. The examples are: DP-DHL's Saloodo (https://www.saloodo.com), DB Schenker's Drive4Schenker (https://d4s.dbschenker.com), and Damco's Twill (https://www.twill.net). On the other hand, they are aware that their business needs to innovate continuously and requires creativity and agility that they, as large corporations, often lack. Therefore, they partner with startups, which are full of ideas, agile, and willing to take risks. Startups enter this cooperation because they lack financial resources and execution procedures, which by contrast are the core competence of large LSPs. The cooperation takes different forms.

UPS, for example, has developed a Strategic Enterprise Fund (corporate venture capital) to invest in startups that are strategically relevant, and help UPS become the leading provider of technologically advanced services in the T&L industry. UPS adopts innovations in the following categories: 3D printing, Artificial Intelligence, automation, collaborative consumption, Internet of Things, smart cities, sustainability, transportation. According to Oliver Wyman (2017, p. 2) in 2016 UPS spent 600 million USD across 24 investments. In March 2019 UPS announced a partnership with Matternet, a startup that builds and operates drone logistics networks for transporting goods on demand, through the air. UPS and Matternet have partnered to deliver medical samples via drones across WakeMed, a major healthcare system in North Carolina. This is the first FAA-sanctioned use of a drone system for routine revenue flights in the USA. The Matternet M2 Drones replace couriers' cars. This collaboration is the latest UPS program to utilize drone flights in support of healthcare logistics. In 2016, UPS partnered with GAVI and Zipline to deliver blood products to remote locations in Rwanda. The Matternet team has already completed more than 3000 flights for healthcare systems in Switzerland.

DP-DHL has also selectively invested in startups that promise short- to mid-term leverage. At the end of 2018 it announced a 21 million USD venture into Resilience360, the supply chain risk management startup with a mobile app for tracking earthquakes, wildfires, cyberattacks, and anything that disrupts supply chain operations. Moreover, DP-DHL established in-house accelerator Start-upLab and joined consortium accelerator Plug-and-Play Supply Chain. Both programs are aimed to support selected startups in searching for innovative solutions to industry challenges. DP-DHL is partnering with startups focused on, e.g., solar energy utilization, digital manufacturing, or inventory management.

DB Schenker promotes an open innovation approach. It engages in different innovation networks such as the International Data Space Association, IDG (International Data Group), HOLM (House of Logistics and Mobility), and the Deutsche Bahn Group, and also partners with startups. A well-known example is the partnership with American freight exchange uShip for developing the Drive4Schenker platform, which aims to improve transport capacity utilization.

While asked about challenges in the LSP-startup cooperation, the LSP interviewees emphasized first, identifying the most promising innovative startups for collaboration and second, choosing the model of collaboration with efficient control mechanisms that do not interfere with the entrepreneurial spirit and innovation culture of the startup. As one LSP reported: "If you want a startup to pursue forward-thinking innovations of strategic importance, you have to leave it some autonomy." That is why standard operative KPIs, such as the return on investment, the degree of customer satisfaction, the revenue increase, savings, the total cost of ownership, or the new competitive advantage in a core business line should not be applied short term to allow for some time for the startup to develop.

The example of UPS shows that corporate venturing is an efficient way to innovate. Even small and mid-size companies can employ this technique to tap into the skillsets, technologies, and ideas of startup companies seeking investment funds. The interviewees told the author that corporate venturing can bring "knowledge returns" far beyond the monetary returns tied to such an investment. They stressed the importance of any, even small, investment and admitted that: "Investing in the startups that you want to learn from ensures that you stay involved with them for the long haul, through the ups and downs. As a result, you learn far more than if you did not invest."

Answering the question of how they process collaborating with startups, a UPS manager explained that UPS is actively engaged with the startups it invests in. At a minimum, a UPS leader sits as an observer on the board of the startup, and sometimes UPS takes a full board seat. Moreover, managing director of the UPS Strategic Enterprise Fund introduces startup leaders to relevant groups and individuals within UPS, so that these groups can partner with the startups on projects, embark on growth initiatives, or just interact informally. UPS people who are involved with the startups become a channel to the rest of the UPS organization, providing a window into how new technologies are evolving.

5 Conclusions

The study shows that the logistics service market is changing, with startups playing an important role in it. On the one hand they are threatening competitors, forcing established players to enhance their innovativeness. On the other hand, they are partners who support established logistics companies in improving their operational efficiency and elevating customer service. In many cases, they can definitely be the springboard that fuels LSPs' innovativeness. Partnering with startups takes different forms. While cooperating, the most important for an LSP is to give a startup not only funding and mentoring, but also the autonomy to develop.

At the theoretical level, the research contributes to enriching logistics studies by collecting, systemizing, and presenting logistics startups themselves, as well as their impact on the logistics service industry's innovativeness. At a managerial level, the findings provide an illustration on how to manage cooperation within corporate venture capital to increase LSPs' innovativeness. The main limitation of the study, which needs to be taken into account when interpreting its findings and conducting future research, is the relatively low number of interviews within the study. Future studies could investigate the LSP-startup cooperation from a startup's perspective.

Acknowledgements This work was supported from Statutory Fund of the SGH Warsaw School of Economics [grant numbers KZiF/S/05/18].

References

1. Riedl, J., Jentzsch, A., Melcher, N.C., Gildemeister, J., Schellong, D., Höfer, C., Wiedenhoff, P.: Why Road Freight Needs to Go Digital—Fast (2018)
2. Wyman, O.: How start-ups are digitalizing logistics? Travel Transp. Logist. (2017)
3. Cichosz, M.: Digitalization and competitiveness in the logistics service industry. E-mentor. **5**(77), 72–82 (2018)
4. Breschi, S., Lassébie, J., Menon, C.: A Portrait of Innovative Start-Ups Across Countries, OECD Science, Technology and Industry Working Papers, 2018/02. OECD Publishing, Paris (2018)
5. Skala, A.: Digital Startups in Transition Economies. Palgrave Pivot, Cham (2019)
6. Blank, S.: Why the lean start-up changes everything. Harv. Bus. Rev. **91**(5), 63–72 (2013)
7. Ries, E.: The Lean Startup: How Today's Entrepreneurs Use Continuous Innovation to Create Radically Successful Businesses. Crown Business, New York (2011)
8. Christensen, C., Raynor, M.: The Innovator's Solution: Creating and Sustaining Successful Growth. Harvard Business Review Press, Boston (2013)
9. Damodaran, A.: Valuing Young, Start-up and Growth Companies: Estimation Issues and Valuation Challenges. Stern School of Business, New York University, New York (2009)

10. Glinka, B., Pasieczny, J.: Tworzenie przedsiębiorstwa: szanse, realizacja, rozwój. Warszawa: Wyd. Naukowe Wydziału Zarządzania Uniwersytetu Warszawskiego (2015)
11. Cieślik, J.: Iluzje innowacyjnej przedsiębiorczości. Kwartalnik Nauk o Przedsiębiorstwie. **3**, 4–16 (2014)
12. Łuczak, K.: Rachunkowość innowacji na przykładzie przedsiębiorstw określanych mianem start-up. Zeszyty Naukowe Uniwersytetu Szczecińskiego. Finanse. Rynki finansowe. Ubezpieczenia. **827**(70), 79–87 (2014)
13. Bursiak, L.: Diagnoza i pozycjonowanie sytuacji finansowej firm start-up oraz spółek z rynku Newconnect w roku 2011. Zeszyty Naukowe Uniwersytetu Szczecińskiego. Finanse, Rynki Finansowe Ubezpieczenia. **59**, 49–58 (2013)
14. Latusek-Jurczak, D.: Firmy szybkiego wzrostu w środowisku niskiego zaufania–badanie empiryczne. Studia Oeconomica Posnaniensia. **5**(9), 37–49 (2017)
15. Kurpjuweit, S., Wagner, S.M., Greil, T.: The rise of logistics startups and their impact on the logistics industry. Conference Paper (2018)
16. Grawe, S.J.: Logistics innovation: a literature-based conceptual framework. Int. J. Logist. Manag. **20**(3), 360–377 (2009)
17. Flint, D.J., Larsson, E., Gammelgaard, B., Mentzer, J.T.: Logistics innovation: a customer value-oriented social process. J. Bus. Logist. **26**(1), 113–147 (2005)
18. Wagner, S.M.: Tapping supplier innovation. J. Supply Chain Manag. **48**(2), 37–52 (2012)
19. Tether, B.S., Tajar, A.: Beyond industry–university links: sourcing knowledge for innovation from consultants, private research organisations and the public science-base. Res. Policy. **37**(6-7), 1079–1095 (2008)
20. Bellingkrodt, S., Wallenburg, C.M.: The role of external relationships for LSP innovativeness: a contingency approach. J. Bus. Logist. **34**(3), 209–221 (2013)
21. Cichosz, M.: Proces innowacji w relacjach operatorów logistycznych z klientami. Gospodarka Materiałowa i Logistyka. **6**, 2–8 (2016)
22. Świtała, M.: Typology of logistics service providers according to the level of innovativeness. Handel Wewnętrzny. **360**(1), 317–333 (2016)
23. Nieströj, K., Świtała, M.: The impact of innovativeness of logistics service providers on their relationships with customers. Gospodarka Materiałowa i Logistyka. **8**, 2–10 (2015)
24. Cichosz, M., Goldsby, T.J., Knemeyer, A.M., Taylor, D.F.: Innovation in logistics outsourcing relationship-in the search of customer satisfaction. LogForum. **13**(2), 209–219 (2017)
25. Wieteska, G., Ocicka, B.: Otwarte innowacje jako źródło wartości w relacjach z dostawcami. In: Rutkowski, K. (ed.) Zarządzanie łańcuchem dostaw w XXI wieku, pp. 129–147. Oficyna Wydawnicza SGH (2016)
26. Wallenburg, C.M., Schäffler, T.: Performance measurement in horizontal LSP cooperation as a field of conflict: the preventive role of collaborative processes. Logist. Res. **9**(1), 7 (2016)
27. Kohler, T.: Corporate accelerators: building bridges between corporations and startups. Bus. Horiz. **59**(3), 347–357 (2016)
28. Weiblen, T., Chesbrough, H.W.: Engaging with startups to enhance corporate innovation. Calif. Manag. Rev. **57**(2), 66–90 (2015)
29. Katila, R., Rosenberger, J.D., Eisenhardt, K.M.: Swimming with sharks: technology ventures, defense mechanisms and corporate relationships. Adm. Sci. Q. **53**(2), 295–332 (2008)
30. Vandaie, R., Zaheer, A.: Surviving bear hugs: firm capability, large partner alliances, and growth. Strateg. Manag. J. **35**(4), 566–577 (2014)
31. Prashantham, S., Kumar, K.: Engaging with startups: MNC perspectives. IIMB Manag. Rev. **31**(4), 407–417 (2019)
32. Cichosz, M.: Otwarte innowacje: technologiczne partnerstwa w branży usług logistycznych. Gospodarka Materiałowa i Logistyka. **4**, 12–22 (2018)
33. Edmondson, A.C., McManus, S.E.: Methodological fit in management field research. Acad. Manag. Rev. **32**(4), 1246–1264 (2007)
34. Yin, R.K.: Case Study Research Design and Methods, 5th edn. Sage, Thousand Oaks, CA (2014)

Modelling the Relations Between the Primary Resources and the Financial Result: Trans Polonia Group Case Study

Andrzej Letkiewicz and Michał Suchanek

Abstract Each entity builds a specific model of functioning, which is related to the selection and configuration of resources, so that the assumed goal can be achieved in given market conditions. This goal in the short term amounts to achieving the planned profit, while in the long term it means competitive advantage and durability of being. This objective is accomplished by shaping the level of used and necessary resources, such as material and personal resources. The article assumes that for transport companies the basic resources that affect the level of profit are means of transport and employment. The thesis of the article is that it is possible to build a model describing the determinants of the use of resources to generate a certain profit by transport companies. The test method of Engle–Granger cointegration was used to build the model. The study was based on data of the Transpolonia entity.

The research has allowed to confirm the usability of the Engel–Granger method to verify the relations between the primary resources of a company and its financial result.

Keywords Transport enterprises · Assets effectiveness · Cointegration test · Allocation of resources · Factors affecting income

1 Introduction

Enterprises have to look for competitive advantages. To make this possible, it is necessary to identify the sources of competitive advantage of a given entity with respect to others. The sources of competitive advantage should be sought in two areas. The first is the market area, while the second is the area of functional processes associated with a unique way and mechanism for configuring the resources

A. Letkiewicz · M. Suchanek (✉)
Faculty of Economics, Chair of Economics and Management of Transportation Companies, University of Gdansk, Gdansk, Poland
e-mail: ekoalt@univ.gda.pl; m.suchanek@ug.edu.pl

M. Suchanek (ed.), *Transport Development Challenges in the 21st Century*, Springer Proceedings in Business and Economics,
https://doi.org/10.1007/978-3-030-50010-8_2

necessary to produce a specific transport product. In cargo road transport, market conditions are impossible to shape by a single enterprise. The high competition prevailing in the market forces companies to seek advantage in the conditions of specialization and acting within narrow market niches. It is on a narrow, niche market that the basis for management efficiency is created in the form of higher unit prices of transport production. This does not absolve transport managers from seeking efficiency within management methods and mechanisms. Each domain of management—production, services, trade—is characterized by a specific way of selecting and using resources to generate incomes in a short period compared with the costs of obtaining them and creating economic efficiency conditions. In the long term, the goal is the accumulation of short-term profits creating conditions for building a lasting ability to maintain competitive advantage. Freight transport is a special example here, as high competition forces enterprises to look for a chance of durability of competitive advantage in the area of configuration of costs of resources used to create a transport service. In this type of activity, the two basic resources are resources in the form of transport means and human resources. Human resources take a form of personnel responsible for acquiring transport work as well as personnel responsible for direct execution. Hence the hypothesis of the article is that it is possible to build a model describing the conditions of using resources to generate a certain profit while ensuring the efficiency of management and durability of competitive advantage by transport companies.

2 Basic Resources in Durability of Competitive Advantage and Efficiency of Transport Enterprises

One of the conditions of the durability of road transport companies is their ability to cope with the turbulences in the environment. These arise in both the macro (global, regional, national) and the micro (industrial, organizational) scale—always, everywhere, and suddenly. Thus, the most important factor that determines the survival capability and sustainability of the economic activity of business entities is the ability to pick up the warning signals while striving to adapt company objectives and functioning mechanisms to the requirements of the environment. Consequently, these entities are obliged to set the directions and methods of operating by identifying or forecasting changes in the environment. This allows business entities to survive even under less favourable future conditions. Only a handful of companies are able to use pre-emptive action to seize all benefits of competitive advantage so as to make the situation more difficult for the competition. Such action is risky, as it requires a serious commitment of resources based on incomplete information. The possible scenarios should take into account potential restrictions on the efficiency and effectiveness of management and the distribution of the various forces that cause and counteract change to ensure the long-term durability of economic activity. This requires introducing changes so that the company would be able to respond to new demands of the external and internal environment [1, p. 69].

Permanent improvement of the functioning of transport enterprises lies in the intangible nature of transport services, which in turn implies an extremely important role of shaping positive employee attitudes in the field of customer service and skills, competence of service sellers, dispatchers, controllers and executive staff-drivers. Therefore, the marketing communication techniques, external and internal, are supporting the shaping of the identity, reputation, image, and brand of the transport company and are playing an important role in the management processes of the transport company. They facilitate building lasting relationships between the company and the environment [2, p. 573], contributing to the sustainability of competitive advantage. The sequence of possible interdependencies is shown in Fig. 1.

Based on a sequence of market transport operator success factors, two elements should be noted. First of all, factors that create income like customer satisfaction because the higher level of the satisfaction builds inclination to pay higher cost of transport services by the client. However, it should be remembered that there are limitations to the increase in prices for transport services, resulting from the mechanism of market equilibrium. Free shaping of prices for services is possible only in a monopoly situation (which is not the case for the Polish road transport market) or in the situation of offering a pioneering service, not offered on the market. However, the possibility of price setting is possible only in the short term because, as it was written in the introduction, competitive advantage requires continual upkeep. Otherwise competitive advantage followers will appear. Therefore, one of the tools

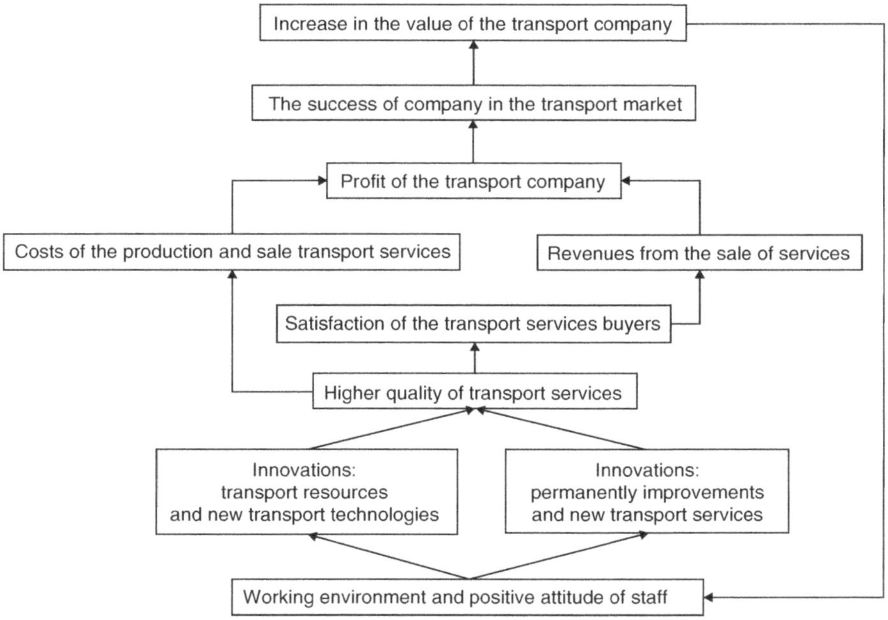

Fig. 1 Factors creating the market success of the transport enterprises. (Source: Own elaboration adapted from [2, p. 574])

to care for and build a competitive advantage is to improve the quality of services by building and modernizing transport potential and induct new transport technologies.

Building and modernizing transport potential and induct new transport technologies changes the costs of service realization. When referring to the costs of transport services, it should be stated that they form part of the issue of management efficiency in general. The general process of building effectiveness is clearly determined by legal and market conditions, which affect the business principles and operating conditions and thus become the basis for individualizing the company's market, social, operating, financial and economic objectives. The degree of their achievement is referred to as the company's efficiency, and the foundation of business durability is the realization of its basic processes, described both by the effort necessary to implement them and their results—in other words, by the input and the output in terms of their productivity, efficiency, profitability and earning capacity (Fig. 2). These factors determine their ability to adapt to the market conditions [4, p. 13].

It is necessary to take into account as many factors related to inputs and outcomes as possible, so income and costs that can be expressed in financial or technical units should be included. In the analysis of economic effectiveness, it is quite easy to take into account only those elements of expenditures and profits that have some value and can be expressed in a monetary form, hence it can be said that the traditional economic calculation is in some respects imperfect because in it it's dif-

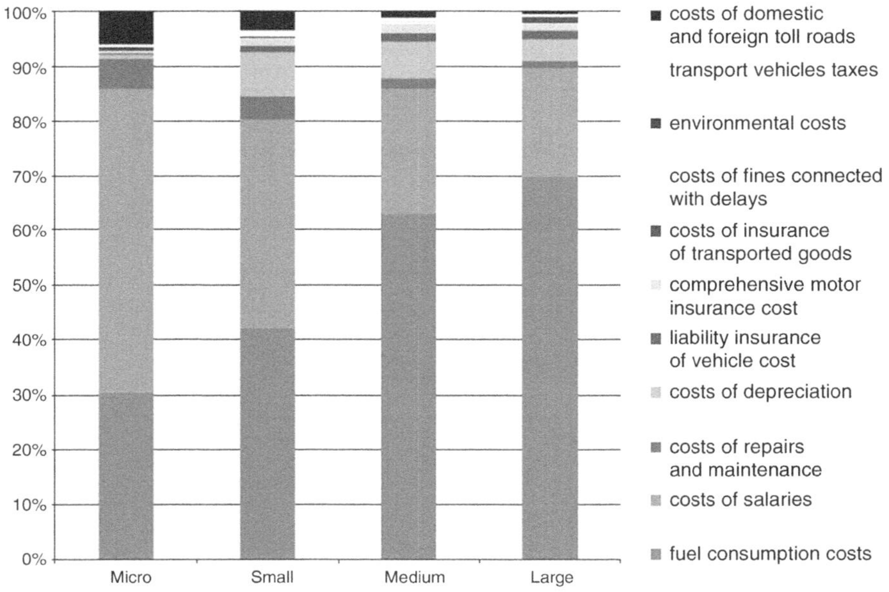

Fig. 2 Costs structure in road transport according to company size (percentage). (Source: Kot [3, p. 392])

ficult to take into account additional, intangible factors related to the business activity [5, p. 1185].

It should be also noted that in most cases in methodological approaches concerning assessment of the financial and economic security of enterprises, systems of partial indicators are proposed. They characterize precisely the financial short-term stability of enterprises, not the level of their financial and economic long-term security. It does not ensure possibility for assessment of effectiveness management process of enterprises and for quantitative and qualitative threat effect assessment of non-measurable factors which come from macro- and mesoenvironment [6, p. 2]. Therefore, the effectiveness of resource use in addition to economic factors is created by non-material factors such as ecological, public and certain social and business norms as well, that cannot be infringed and are not included in effectiveness models. At this point it should be noted that these effectiveness norms are of particular importance for service enterprises, for transport enterprises in particular, because costs of respecting them are a part of the financial model of enterprise functioning and it becomes an element of the relation between an income generated as a result of competitive advantage and the financial result of management processes.

There are studies that present the cost structure in Polish road transport companies. It can be recalled that S. Kot's research shows that there are high differences in cost structures in road transport companies (Fig. 2) but the study takes into account only direct costs such as: costs of domestic and foreign toll roads, transport vehicles taxes, environmental costs, costs of fines connected with delays, costs of insurance of transported goods, comprehensive motor insurance cost, liability insurance of vehicle cost, costs of depreciation, costs of repairs and maintenance, costs of salaries, fuel consumption costs. It can be confirmed that larger companies cut the costs inside the company better than small enterprises. The image of the cost structure allows to draw two conclusions—the first states that the costs of remuneration constitute from 20 to 35% of costs, and secondly that large and medium-sized enterprises transfer non-measurable costs into indirect costs, unrelated to services realization, which affects the efficiency of management and requires analysis by another method.

3 Trans Polonia Group as a Subject and Research Method

Trans Polonia Group is a specialized service provider in the field of transport and logistics of: liquid fuels, petrol, diesel, and liquified petroleum gas (LPG); liquid chemicals; liquid bitumen; liquid food products. The Group operates across Europe leading its operational management from the Polish offices in Tczew and Plock as well as foreign offices in Holland, France, Germany and Spain. An international team of Trans Polonia Group employees consists of the top-notch specialists whose competence is a result of undertaken trainings confirmed by certificates evidence and many years of experience in service of the largest petrochemical plants. Among the clients of Trans Polonia are the largest companies in the petrochemical industry.

Table 1 Value of vehicles and employment Trans Polonia Group

Year	Net profit	Value of vehicles	Employment in persons
	(in thous. PLN)		
2011	4283	11,596	19
2012	2083	18,267	21
2013	3808	26,753	26
2014	2616	34,091	38
2015	457	27,571	38
2016	8217	53,000	536
2017	10,462	60,567	505

Source: Own elaboration from period reports Trans Polonia Group (https://transpolonia.com, 05.03.2019 [7])

Cooperation in this highly specialized market is based primarily on the high quality of services offered, the reputation of the supplier and the ability to build mutual trust. Trans Polonia meets the highest standards in the industry from the ADR convention through the ISO 9001:2015 to SQAS. Trans Polonia successively invests in modern equipment used for transport of liquid raw materials and fuel, expanding its fleet to more than 500 tanker sets. The Group has developed a very flexible model of international business (https://transpolonia.com, 05.03.2019 [7]). The basic data characterizing the company's assets in the form of the value of vehicles, employment and net profit in the period 2011–2017 are presented in Table 1.

4 Material and Methods

In order to prove the hypothesis that the financial result is directly caused by the value of motor vehicles on hand in a company, data on these two variables have been collected for motor vehicles as well as two other companies which serve as the control group: Balticon and ATC CARGO (Table 2).

We want to prove that the value of vehicles affects the financial result of the company. Thus the method of choice is vector autoregression (VAR) followed by a Granger causality test. VAR models are time series models in which there is an assumption that all the variables are endogenic, there are no artificial constraints regarding the number of variables in the model. They are widely used due to the fact that they provide an easy possibility of interpretation and they are free from too many theoretical assumptions [8]. The following Engle–Granger test is used to verify if there is a causal relationship between variables, i.e. if an introduction of a lagged variable as a cause of a different variable which is the effect improves the prediction value of a model as a whole. The test verifies the null hypothesis stating that there is no causality after a VAR model has been estimated [9].

Table 2 Variables used in the model (years 2011–2017)

Variable	Mean	Standard deviation	Min	Max
TP_NetPR	4560.857	3549.619	457	10,462
TP_Vehi	33,120.71	17,817.22	11,596	60,567
BC_NetPR	−329.778	3486.163	−8153.29	1617.957
BC_Vehi	6240.891	2238.87	3895.044	10,569.49
AT_NetPR	1114.447	750.73	334.955	2500.382
AT_Vehi	1421.635	427.3178	1058.696	2167.224

Source: Own elaboration based on financial reports of the companies (05.03.2019)

Table 3 VAR models and Engle–Granger tests

Values	Trans Polonia model	Balticon model	ATC CARGO model
Lagged financial result parameter	0.12	−0.27	0.09
p-value	0.75	0.67	0.86
Lagged motor vehicles parameter	0.21	0.03	−0.51
p-value	0.03	0.97	0.59
R-squared	0.86	0.07	0.36
AIC	18.24	20.06	15.54
Cointegration – Vehicle p-value	0.02	Not done	Not done
Cointegration – Time p-value	0.08	Not done	Not done
Granger test p-value	0.01	Not done	Not done

Source: Own elaboration

5 Results

Three VAR models were estimated for all the three companies with the net profit treated as the endogenic variable and the value of motor vehicles treated as the exogenous variable. Parameters for individual variables were estimated along with R-squared and Akaike Information Criterion (AIC). For the formally sound models, Engle–Granger test was performed to verify the causal relation between the variables. The results are presented in Table 3.

We've been able to generate a formally sound VAR model only in case of Trans Polonia. The results indicate that while the financial result of the company is not directly affected by the value of the financial results in the preceding years (lags up to 4 years were tested), it is significantly affected by the value of motor vehicles on hand. In fact, an increase of the value of motor vehicles by 1 PLN resulted in an increase of the financial results by 0.21 PLN in the following year. This has been further proven by the Engle–Granger test which proved that an introduction of the causal effect improves the prediction power of the model. In fact, in case of Trans Polonia, the value of motor vehicles is one of the main drivers of the growth of the financial results. This has not been the case for the other two companies, nor has it been true for the number of employees which was also tested as a potential exogenous variable.

6 Conclusions

The results of the model prove that it is possible to create a model of relation between the financial result and the value of primary resources in a company, in this case the motor vehicles. There is, however, a number of reasons while this might not always be the case. First of all, the model turned out to be statistically significant and formally correct only in one of the three cases analysed, while the models for two other companies were impossible to estimate. There might be a number of reasons for that. Trans Polonia is the biggest companies out of the three. It is also a well-established company with a long history of presence at the stock market during which it has mostly steadily grown. Second of all, it is a company which generates most of its revenues based on contracts with public companies, mostly energy companies, which process crude oil. This means that Trans Polonia has a stable revenue source. What's more unlike the two other companies, it has had the same manager for over 20 years and its style of governance remains more or less the same. This means that the management has had enough time to develop and mature internal regulations which allow the company to react better to the market changes. This means that it is not as affected by the market turbulence as the two other companies might be. Lastly, we have chosen transport companies as a basis of the construction of the model of relations between the financial results and the primary resources because in a transport company, especially a road cargo transport company, there is an obvious group of assets which can be described as the primary resources, i.e. the motor vehicles. In other companies, it might be harder to distinguish the assets which are primarily responsible for the generation of revenue and in turn income. Still, the example proves that such a model is possible, especially for experienced and mature companies. Further research in this matter should concentrate on creating a large group of models for a representative part of the market, preferably chosen in a random way so as to satisfy the formal demands of econometric modelling and statistical inference. At this point, the hypothesis of the paper can only be said to be partially confirmed.

References

1. Letkiewicz, A., Suchanek, M.: Systemic conditions of efficiency and durability of road transport enterprises. In: Letkiewicz, A., Suchanek, M. (eds.) Determinants of Functional Effectiveness and Durability of Road Transport Enterprises. Research Journal of the University of Gdańsk, Transport Economics and Logistics, vol. 73. Gdańsk University Press, Gdańsk (2017)
2. Wojewódzka-Król, K., Załoga, E. (eds.): Transport. Nowe wyzwania. WN PWN, Warszawa (2016)
3. Kot, S.: Cost structure in relation to the size of road transport enterprises. Promet Traffic Transp. **27**(5), (2015)
4. Fryca-Knop, J., Letkiewicz, A., Majecka, B.: Functional efficiency and stability of road transport enterprises. In: Letkiewicz, A., Suchanek, M. (eds.) Determinants of Functional Effectiveness and Durability of Road Transport Enterprises. Research Journal of the University of Gdańsk, Transport Economics and Logistics, vol. 73. Gdańsk University Press, Gdańsk (2017)

5. Drop, N.: Ocena efektywności inwestycji w środki transportu na przykładzie przedsiębiorstwa z branży transport-spedycja-logistyka (TSL). Autobusy Technika, Eksploatacja, Systemy Transportowe, nr 6/2018
6. Novikov, A.O., Novikova, M.M.: Modelling of financial and economic security of transport enterprises based on factor analysis. Sci. Transp. Progr. **6**(54), 7–22 (2014)
7. https://transpolonia.com. Accessed 5 Mar 2019
8. Johansen, S.: Modelling of cointegration in the vector autoregressive model. Econ. Model. **17**(3), 359–373 (2000)
9. Bilgili, F.: Stationarity and cointegration tests: comparison of Engle-Granger and Johansen methodologies. Erciyes Üniversitesi İktisadi ve İdari Bilimler Fakültesi Dergisi. **13**, 131–141 (1998)

Changes in Structure and Spatial Distribution of Heavy Goods Vehicles Traffic in Poland in 2005–2015, with Particular Emphasis on Border Traffic

Piotr Rosik, Tomasz Komornicki, Sławomir Goliszek, Patryk Duma, and Barbara Szejgiec-Kolenda

Abstract The purpose of the chapter is to show the processes which have an impact on the size, structure, and spatial distribution of heavy goods vehicles traffic over the network of national and voivodeship roads in 2005–2015. A particular attention was paid to the border areas and border crossings. The analysis was performed on the basis of a number of databases, including Traffic Census as of 2005, 2010, and 2015, Polish Border Guard traffic database, the modal structure of export and import at the poviat (county) level (Central Statistical Office—GUS), and the National Traffic Survey (Central Board of National Roads and Motorways—GDDKiA). The border (external) traffic was divided into origin-destination traffic and transit traffic.

Keywords HGV traffic · Border areas · Border traffic · Export flows · Import flows · Transit traffic

P. Rosik (✉) · T. Komornicki · S. Goliszek · P. Duma · B. Szejgiec-Kolenda
Department of Spatial Organization, Institute of Geography and Spatial Organization, Polish Academy of Sciences, Warsaw, Poland
e-mail: rosik@twarda.pan.pl; t.komorn@twarda.pan.pl; sgoliszek@twarda.pan.pl; p.duma@twarda.pan.pl; b.szejgiec@twarda.pan.pl

M. Suchanek (ed.), *Transport Development Challenges in the 21st Century*, Springer Proceedings in Business and Economics,
https://doi.org/10.1007/978-3-030-50010-8_3

1 Introduction: Purpose of the Study

Heavy goods vehicle road traffic modeling at the national level in Poland has not been hitherto the subject of in-depth spatial analysis. This is mainly due to the lack of comprehensive traffic studies, covering the entire territory at the level of possibly small transport regions. Most of the analyses have been conducted for the individual traffic at the level of cities or agglomerations and in the recent years—also voivodeships (i.e., provinces, administrative NTS 2 regions, see [1], and an exceptional work devoted to the modeling of truck traffic at the level of Małopolska province in [2]). At the national level the so-called [3] was developed, based on the Traffic Census of 2005 with several dozen control points, only for the national road network. As a result, the traffic estimations, due to lack of consideration of such crucial aspects as trade and concentration of economic potential in metropolises, were largely detached from broader social and economic context.

This chapter attempts to model heavy goods vehicles road traffic by means of secondary statistical data. These data present trade flows with distinction of the means of transport. Matrix data for trade from poviats to foreign countries were accounted for regarding three HGV types. These data illustrate the actual relations (flows) of each unit (poviat) with border crossing points. Data concern three HGV types, which are: light vehicles, heavy goods vehicles without trailers, and heavy goods vehicles with trailers.

The subject of the study was constituted by the spatial diversification of truck traffic and, consequently, the intensity of truck traffic on the networks of national and provincial roads. The chapter presents the division of lorries into three categories of vehicles: light trucks with a maximum total weight (GVW) below 3.5 t (so-called delivery vans), lorries without trailers, and lorries with trailers and semi-trailer tractors. This division is consistent with the division defined as a part of the Road Traffic Census.

The main objective of this study is to develop the modeling of external traffic of heavy goods vehicles for the entire territory of Poland at a detailed spatial scale (involving a large number of transport regions) using statistical data indicating local conditions related to spatial and socio-economic structure along with functional relations. In this chapter, we focus on the processes, which bear an impact on the size, structure, and spatial distribution of heavy goods vehicles traffic over the network of national and provincial roads in 2005–2015. A particular attention was paid to the border areas and border crossings.

The assumed cognitive goal of the study reported is to identify the factors, affecting the distribution and intensity of heavy goods vehicle traffic. The methodological objective intends to propose a research methodology enabling forecasting traffic for the entire country over the network of national and regional roads in the form of a model, acting as an additional tool facilitating application of the available secondary statistics in the form of matrix data. Authors' own model for traffic speed has been implemented, including a number of factors affecting the speed of heavy goods vehicles. For the purpose of calculations, the VISUM software has been employed.

The research was carried out under the scientific grant entitled: *Monitoring changes of traffic-generating potentials and determinants of the distribution of freight traffic on the network of Polish roads in the years 2005–2015 (HGV-Monit)*, conducted at the Institute of Geography and Spatial Organization Polish Academy of Sciences (IGSO PAS further on) in 2014–2019. The project was financed from the National Science Centre funds, granted on the basis of Decision No. DEC-2014/13/B/HS4/03351.

2 Scope of the Study

The time range of the study was determined for the years of 2005–2015, on account of conducting the General Traffic Measurement (GPR) on a network of national and regional roads every 5 years, meaning the years 2005, 2010, and 2015. A considerable part of statistical material and data concerning the matrix data for trade also refers to the years 2005, 2010, and 2015. Monitoring studies, i.e., in dynamic terms, are also very rare and are limited at the national level to accessibility only [4, 5]. Thus, it is the first such type of study referring to heavy goods traffic flows, taking into account the dynamics of the phenomenon in the years of 2005–2015.

The analysis was conducted on the basis of a number of databases, including the Road Traffic Census as of 2005, 2010, and 2015, Polish Border Guard traffic database, the modal structure of export and import at the poviat level, provided by the Central Statistical Office (GUS), and the National Traffic Survey (Central Board of National Roads and Motorways—GDDKiA). The border (external) traffic was divided into origin-destination traffic (based on export and import county-level database) and transit traffic.

The spatial scope of the study concerns the road network on the entire territory of Poland. This concerns mainly the network of national and provincial roads, but also the more important sections of local roads, i.e., county and municipal roads, were taken into account [6, 7]. The reference point for the analysis (GPR2005, GPR2010, and GPR2015) was the network of national and provincial roads along which the Road Traffic Census was executed. This means that the entire project was carried out for the road network observed at three points of time, i.e., at the end of 2005, 2010, and 2015, respectively.

The majority of the large traffic models that have been recently developed in Poland refer to the number of transport zones ranging between roughly 200 and 300 [8]. Nevertheless, the impact of data aggregation upon the results in transport studies is of great importance [9]. In this study, ten times more—exactly 2321—transport regions defined at the municipal level have been considered [10], although we focus in this chapter on the part of the study, devoted to the external traffic, where in general, 335 county (poviat) seats were taken into account. In addition, the external traffic of heavy goods vehicles at 48 border points (including two seaports, those of Świnoujście and Gdynia) was taken into account for the purposes of analysis.

In the implemented heavy goods vehicles traffic speed model, the speed on 13 road categories (from local roads to motorways) was defined with the use of an original three-stage procedure. Subsequently, at the fourth stage of the procedure the impact of traffic intensity on the speed change is considered. Application of an alternative approach for determining traffic speed involves indirect consideration of other (apart from traffic intensity) speed limitations. The main source of data regarding the speed of heavy goods vehicles traffic in 2005, 2010, and 2015 was the traffic speed model developed by the IGSO PAS team for the needs of isochrone analysis and potential accessibility modeling [5, 10–13].

This model takes into account factors that are seldom considered or get even neglected in standard engineering models, while strongly affecting driving conditions and thus traffic speed. Hence, the estimated traffic speed is not "unhampered," and is intended to approximate the average possible speed, taking into account the provisions of the Highway Code, technical and functional parameters of roads and traffic conditions [12, 14]. Therefore, the heavy goods vehicle speed model has been developed assuming the impact of three variables on vehicle speed, i.e., population number in a 5 km buffer around given road section, built-up area density, and land relief.

3 Geopolitical Conditions of Heavy Goods Vehicles External Traffic

In international freight traffic, legal barriers include customs, bans on the import and export of selected goods, and phyto-sanitary and veterinary restrictions. Operators are required to obtain appropriate licenses and permits, and to comply with the rules of heavy goods vehicles in the given country (e.g., enforcement of the provisions on the technical condition of vehicles and their maximum payload, the

Table 1 Selected geopolitical events affecting the international traffic of heavy vehicles on the territory of Poland in 2005–2015

Year	Geopolitical event
2003	Visa introduction for Russians, Belarusians, and Ukrainians
2004	The accession of Poland and of Poland's neighbors: the Czech Republic, Slovakia and Lithuania, as well as Hungary, Slovenia, Latvia, Estonia, Cyprus, and Malta to the European Union
2005–2007	Russian embargo on food
2007	Accession of Romania and Bulgaria to the European Union
2007	The accession of Poland and other countries that joined the EU in 2004 (except Cyprus) to the Schengen area
2011	Russian embargo on fresh fruits
2014–	Annexation of Crimea and conflict in eastern Ukraine
2014–	Russian embargo on meat and fruits and vegetables from Poland

obligation to buy vignettes, traffic limitations during holidays and weekends, etc.) [10]. At the international level, legal conditions in the form of international trade and transport agreements as well as agreements between carriers and forwarders and other regulations are important in the context of geopolitical conditions.

In the context of customs and trade barriers, the key date was Poland's accession to the European Union in 2004 and, to a lesser extent, 3 years later—the accession to the Schengen area (see Table 1). During the analyzed period, along with the accession of—in particular—Poland, Lithuania, the Czech Republic, and Slovakia to the EU, the customs borders between these countries and the remaining countries of the European Union were abolished. Trading within the European Union is not covered by the customs duty, and the SAD document (customs declaration) is replaced by the INTRASTAT declaration.

The entry into the Schengen zone had some consequences in terms of further reduction of the travel time of trucks in international traffic, reduction of the cost of transport within the Schengen area, and, in connection with this—further increase of traffic. The impact of the removal of border barriers on the volume of truck traffic can be considered on two levels:

(a) Intensification of international trade within the European Union, resulting directly from the opening of markets, which is manifested in export and import statistics
(b) Reduction of transaction costs, related to crossing of the border and customs, phyto-sanitary and passport clearance, which is expressed through savings in the time of work of drivers, as well as lack of handling costs (Customs Agencies), which in turn allows for intensification of transport

In the context of the eastern border, after 2000 a renewed increase (after the Russian crisis of 1998) in the traffic on all border sections was observed. The slowest was at the border with Belarus, where only in 2007 the level of traffic from 1997 was attained. The accession of Poland and the Baltic States to the European Union in 2004 caused a spectacular increase in the Polish-Lithuanian movement in 2004–2006, accompanied by (though not on the same scale) decline of traffic on the Polish-Belarusian border. Since 2006, more trucks have been crossing the border with Ukraine than with Belarus. Meanwhile, Budzisko, for which the data ends in 2007, when Poland became part of the Schengen zone, became one of the most heavily charged Polish commodity border crossing. One of the reasons was the transfer of a part of transit from and to Russia to the direction through Latvia and Lithuania, which gave the opportunity of avoiding double border control at the Belarusian borders. This coincided with the growing share of Lithuanian transport companies in servicing the transit market. Traffic across the border with Ukraine also grew quickly. Poland's membership in the Schengen area (since 2007) did not significantly affect the upward trend of traffic on the border with Belarus and Ukraine. In 2009, we observe the collapse of the Polish-Ukrainian traffic, clearly related to the global economic crisis. On the Belarusian border, there was only a temporary stagnation, while on the Russian side, the downward trend is continuing.

After a short break, the traffic across the border with Belarus again exceeded the Polish-Ukrainian movement.

The following years (after 2009, including the year 2010, a key year from the point of view of this study) were characterized by a rapid growth of road freight transport between the European Union (including Poland) and Eastern Europe. These movements took place through various transport corridors. The Lithuanian corridor maintained its key role, but again the importance of the routes crossing Belarus was strengthened. The exchange with Ukraine also resulted in the growth of the heavy goods vehicle traffic. In 2010–2013, all border sections experienced a significant increase in heavy traffic. In 2013, the traffic across the Polish section of the EU external border (the borders with Ukraine, Belarus, and the Kaliningrad region of the Russian Federation) exceeded, in both directions, 2.0 million heavy goods vehicles.

The positive trend described above was interrupted by the events of 2014, namely the annexation of Crimea and the conflict in eastern Ukraine, introduction of sanctions against Russia by the European Union and, as a consequence, Russian embargo on selected products of the Polish food industry. All in all, the traffic of cars across the eastern border decreased by almost 8%, which we can carefully consider as the result of geopolitical turmoil. The number of crossings of the Russian and Belarusian borders was still higher in 2015 than the one recorded in 2009, but in the case of the Ukrainian border, we can talk about a similar level. The year 2015, as compared to later years, when the number of crossings through the Belarusian and Ukrainian borders increased again, is exceptional and can be considered as the one in which the truck traffic collapsed across the eastern border.

4 Methodology: Procedure

Traffic modeling is a frequent procedure, which is primarily performed for cities and agglomerations for commuting see [14, 15], for education [16, 17], and for tourism [18, 19]. However, according to the best knowledge of the present authors, this is the first study devoted to freight transport for the whole network of national and provincial roads, in which special emphasis is placed on the analysis of external traffic.

In general, at the level of national models, external transport origins and destinations can be assigned to: (1) regions or countries neighboring upon Poland (e.g., the model prepared by PKP PLK railway company), (2) border crossings (the national model of GDDKiA). This study uses both of these possible solutions in a sequential system. The reason for this approach is the fact that the external traffic model was based on several different, but still relatively coherent, sources of data, namely:

1. **Export and import database** at the poviat level for the years 2005, 2010, and 2015, broken down into the directions of export and import (countries), transport modes (including road traffic, i.e., goods transported by truck), and description

of the goods; the database gives the possibility of assigning a given traffic flow to the country of origin/destination, e.g., to Russia, without the direct information on the border crossing or the section of the border through which the goods crossed the Polish border; the database gives the opportunity to estimate the origin-destination traffic.
2. **Border traffic database**, concerning the road segments leading to border crossings (based on the Traffic Censuses of 2005, 2010, and 2015).
3. Truck traffic database according to the **database of the Border Guard**; database for all border crossings as of 2005 and for crossings at the eastern border (boundaries with Russia, Belarus, and Ukraine, but not with Lithuania) for 2010 and 2015.
4. The database of surveys carried out at selected border crossings in 2005 as a part of work on the **National Traffic Model**—this database gives the prerequisites for estimating the size and orientation of transit traffic in 2005.

In order to match the databases used in the study, the following steps were taken in the research procedure:

1. **Assignment of the GPR value**. Each of the border crossings was assigned the traffic volume for the road segments next to the border, according to the GPR 2005, 2010, and 2015 according to vehicle types, i.e.,
   (a) LGV—vans and vans with trailers
   (b) HGV—rigid HGV (trucks without trailers)
   (c) TIR—drawbar trailers and articulated HGV (trucks with trailers).
2. The **calculation of the transit traffic flows**. The transit movement calculation for 2005 was made using the database from the National Traffic Model, in which drivers were surveyed, in particular those of the heavy goods vehicles at borders. Transit traffic flows for 2005 were calculated (TIR-TRANSIT) and for each of the border crossings the sum of transit traffic was subtracted from the total sum of lorries with trailers (a simplification was applied, namely—it was assumed that the transit is carried out only using lorries with trailers). In this way, the share of transit in the movement of trucks with trailers at individual border crossings was calculated. Then, due to the lack of other data sources, it was assumed that these shares are fixed for the years 2005–2015.
3. Calculation of the **origin-destination traffic** for trucks with trailers, based on the total movement of this type of trucks, excluding the transit movement. Then, the export and import database at the poviat level for the years 2005, 2010, and 2015 was broken down into directions and transport modes, thus giving an indirect possibility of assigning a given traffic flow to the border section on the basis of specific traffic flow assumptions. It has been assumed that the analysis of origin-destination traffic concerns land borders in road traffic (carried out by means of HGV transport) and maritime border in maritime transport. For the land borders it was found that:

(a) Traffic to Germany and other countries of Western Europe (excluding Austria and Italy) and to all non-European countries is carried out across the German border.
(b) Traffic to the Czech Republic, Italy, Austria, Hungary, and other countries of south-eastern Europe (to the East of Italy and Austria), as well as to Turkey, is carried out via the Czech border.
(c) Traffic to Slovakia takes place through the Slovakian border.
(d) Traffic to Lithuania and Latvia, as well as to Estonia and Finland, and 40% of traffic to Russia is carried out through the Lithuanian border.
(e) Traffic to Belarus, to Kazakhstan and 40% of traffic to Russia takes place through the Belarusian border.
(f) Traffic to Ukraine and Moldova goes through the Ukrainian border.
(g) Traffic to Russia is in 20% carried out across the Russian border (border with the Kaliningrad region).

5 Results: External Traffic

Total HGV movement at the external borders of Poland in the period 2005–2015 increased by 81.6% (from 42.3 thousand to 76.7 thousand vehicles a day). The increase in truck traffic at the borders was therefore more than twice as high as the analogous indicator for the overall national road system in Poland (increase by 30.2%).

The largest share in the increase was observed for trucks with trailers (TIR) (increase of almost 119%), which in 2005 accounted for around 54% of truck traffic across the state borders of Poland, and in 2015 for over 65% of this traffic. Compared to traffic for the entire national road network in Poland, it can be seen that the traffic of heavy goods vehicles with trailers has a correspondingly higher share in the traffic at the state borders (for Poland the share increased from 45% in 2005 to 56% in 2015).

At the same time, the share of light trucks (LGV) and trucks without trailers (HGV) decreased (from 34% to 27% and from 12% to 8% of the HGV traffic at the borders, respectively). These declines correspond to the trends observed for the national road network.

The model distribution of external traffic (Fig. 1) is a resultant of transit movement through Poland and the intensity of Polish foreign trade (export + import). It has been demonstrated that the recorded deconcentration of trade (especially exports) has been reflected through large flows of heavy goods traffic towards and from some smaller cities (e.g., the headquarters of Special Economic Zones and/or large foreign investments).

In transit traffic, new directions and connections appeared, in particular between Budzisko on the border with Lithuania and the border with the Czech Republic (including the new border crossing point in Gorzyczki), as well as the border with

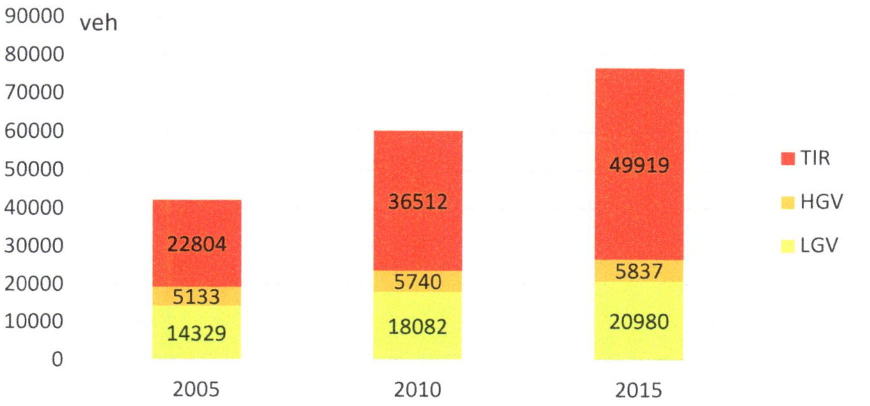

Fig. 1 Total HGV traffic at the external borders of Poland 2005–2010–2015 according to the GPR2005/2010/2015 database (only land crossings); LGV—vans and vans with trailers; HGV—rigid HGV (trucks without trailers); TIR—drawbar trailers and articulated HGV (trucks with trailers)

Slovakia (the crossing point in Barwinek). However, the dominant transit movement direction is still the one between the border crossings (border points) in Świecko and Budzisko. The relative weakening in terms of total movement, compared to other border sections, of the Ukrainian border results in its lesser importance in transit traffic.

Changes in the directions of transit traffic indicate a relatively more pronounced role of the diagonal movement of vehicles (north-east—south-west), compared to the traditional east-west direction. This proves indirectly the growing role of relations with Eastern Europe, generated in the countries of South-West Europe. Likewise, it also points out the growing role of Polish-Czech road infrastructure in the overall European transport system. In addition, transit traffic is naturally more spatially concentrated than the traffic generated by Polish trade relations.

The character and the spatial distribution of trade relations with neighboring countries also influence the distribution and the structure of the movement. As previous studies on the regional structure of Polish exports demonstrate [20], economic contacts with some countries have a strong cross-border dimension. This is especially true for Ukraine, but also for Lithuania and the Czech Republic. This results in intense movement over relatively short distances. In this situation, light commercial vehicles are especially made use of.

The distribution of external traffic in the years 2005–2015 is more or less stable and depends largely on changes in the road network in Poland, and also, although to a lesser extent, on the emergence of new directions in foreign trade, realized with the use of HGV transport. In the years of 2005–2015, the theoretical distribution of external traffic shows changes in central Poland (Fig. 2), related to the completion of subsequent segments of the A1 and A2 motorways, as well as the completion of the S8 expressway, linking Wrocław and Łódź. In addition to changes on these road

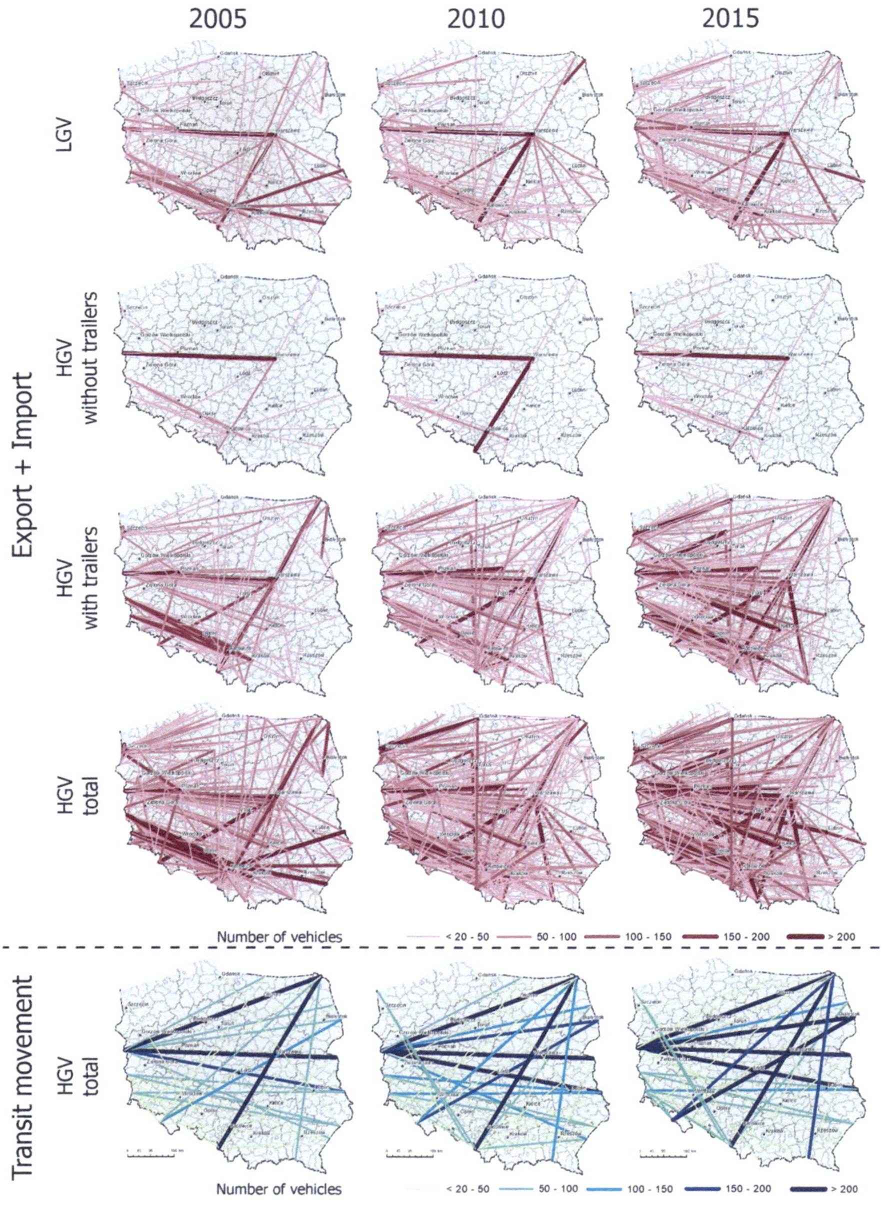

Fig. 2 Theoretical distribution of origin-destination external traffic (based on data on export and import) divided into three types of trucks and external transit traffic; in 2005, 2010, and 2015

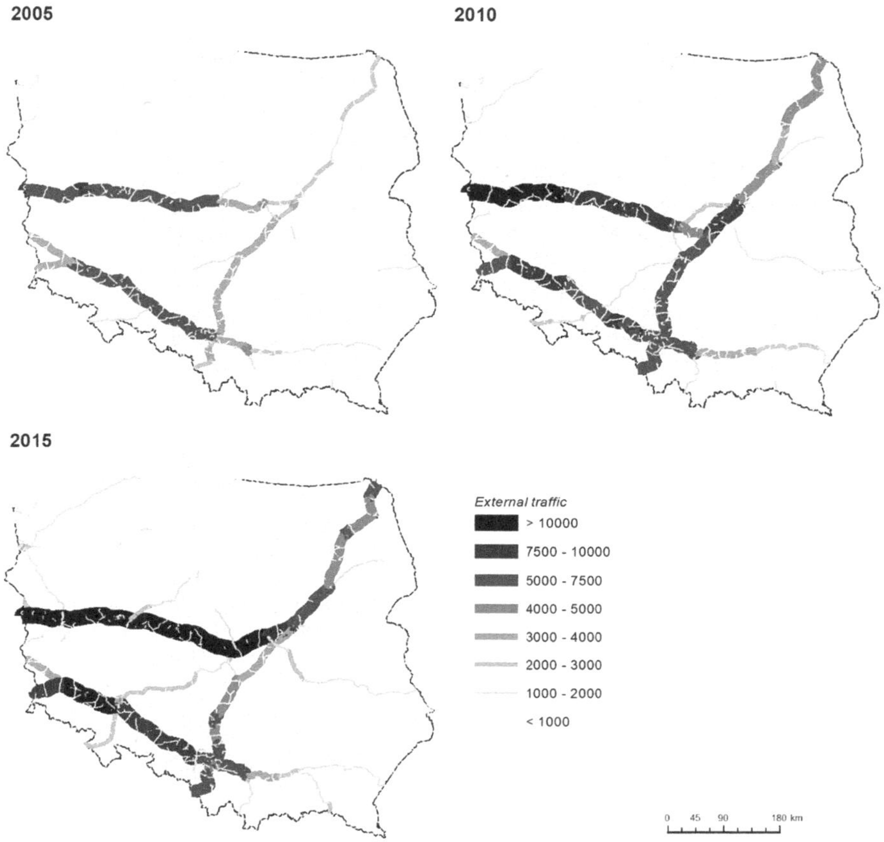

Fig. 3 Theoretical traffic intensity of external traffic (export + import + transit movement) on the national and provincial road network in 2005, 2010, and 2015

segments, in 2015, the intensification of external traffic on the national roads nos. 6 and 5 and over the connections to Slovakia (to Barwinek and Chyżne) can be observed. External traffic to the border with Lithuania in Budzisko, after a period of intensive growth in 2005–2010, has been subject to some stabilization (Fig. 3).

6 Conclusions

Poland's accession to the European Union in 2004, followed by the entry into the so-called Schengen zone determined geopolitically, as a result of the reduction of formal and legal barriers and the elimination of waiting times at the state borders, increased traffic between Poland and other countries belonging to the Schengen area. On the eastern border, in the context of the distribution of truck traffic to particular border sections, the concentration of transit traffic from Russia on the routes through Latvia, Lithuania, and the Via Baltica route to Warsaw and Germany continued in the period after the financial crisis of 2009. At the same time, the intensity of traffic at the Polish-Belarusian crossings gradually increased, in particular in Kukuryki, which can be interpreted as a slow return of transit to the direction of Minsk-Warsaw. The traffic from Ukraine took place only through the direct border with this country and gradually moved again onto the route through Warsaw and the A2 motorway.

Heavy goods vehicles traffic is concentrated on national roads, mainly on dual carriageways, An important role in the analyzed period was played by the construction of the subsequent segments of the network, which results in traffic shifts (e.g., from A2 and A4 motorways to the S8 expressway between Wrocław and Łódź). In the period 2005–2015, the distribution of traffic was the result of two overlapping trends: (a) the general concentration of traffic on the main roads, in particular motorways; (b) deconcentration within the network of motorways and expressways, connected with the expansion of this network. In addition, there are visible traffic shifts, connected with covering of the subsequent segments with the viaTOLL network. It was found that the role of road infrastructure in stimulating traffic patterns is becoming more and more visible as the differences in its quality increase (investment-related progress). The study allows identification of road segments of particular importance for international transport. Some of them are not motorways nor expressways. Others are not even planned as such routes. An example may be provided by the roads serving the southern part of the Podkarpackie province, as well as the route from Wrocław through the Klodzko Valley to the Czech Republic. In addition, based on the model traffic distribution, it is also possible to identify the road nodes overloaded with heavy vehicle traffic. The node of Warsaw belongs among them first of all.

On this basis, it is possible to propose the model stages of changes in the distribution of heavy external traffic on the expanding transport infrastructure. The following stages have been distinguished:

- *Stage 1*. Infrastructure poorly developed. Small length of dual carriageways. Differences in spatial accessibility depend on the distribution of the masses (population, GDP). Traffic is a simple response to demand from freight (with international disruptions).
- *Stage 2*. Economic development takes place under similar conditions as during the stage 1. The traffic is growing and it starts to move or "escape" to more routes. The accessibility pattern is similar, but its real level decreases, due to congestion.
- *Stage 3*. In response to the development, the investment process begins (construction of motorways, expressways, and bypasses) and fees are introduced. With the progress of these works, polarization of accessibility and further scattering of heavy traffic takes place, seeking the most convenient, fastest, and/or cheapest routes.

- *Stage 4*. The most important investments are completed. The distribution of modern infrastructure becomes relatively even. The toll system covers most of the main roads, including the "escape routes." Again, the polarization of accessibility decreases, which also contributes to the overall accessibility improvement (especially in peripheral zones). Movement is again a response to demand. Its concentration on the new infrastructure follows.
- *Stage 5*. A definite improvement in the accessibility of some centers (especially second-order centers) results in their economic development and the generation of new traffic. Infrastructure may turn out to be insufficient again. It is necessary to densify the network or to extend routes. Locally, deconcentration may appear again.

Summing up, the study showed that the distribution, structure, and dynamics of external traffic of lorries depend first of all on: (a) the distribution of economic (in particular export) activity in the country; (b) the quality of the modernized infrastructure, (c) the nature of bilateral economic relations with neighboring countries; (d) a pan-European system of economic ties (including, in particular, those between Eastern Europe and Western EU countries); (e) spatial distribution of road tolls. The proposed methodology may in the future become fundamental for establishing the principles for forecasting external heavy goods vehicles traffic for the entire national road network (by the adequate consideration of specific regional conditions). Knowledge of regional and local determinants is an opportunity for territorialization of transport policy [21]. The role of the developed models and the meaning of the further more detailed research in this field is important both for large agglomerations and for the peripheries. This may be, for example, a hint for the distribution of funds between investments on national and regional roads. It is recommended to conduct extensive surveys, concerning, in a particular manner, the heavy goods vehicles traffic at the national level, which will take into account different aspects of both external and internal HGV traffic, in particular in an era of automated and autonomous vehicles [22].

References

1. Kulpa, T., Szarata, A.: Development of the transport model for the Masovian Voivodeship. In: Contemporary Challenges of Transport Systems and Traffic Engineerin. Lecture Notes in Networks and Systems, vol. 2, pp. 193–204. Springer, Heidelberg (2017)
2. Kulpa, T.: Modelowanie potencjałów ruchotwórczych w drogowych przewozach ładunków w skali regionu. PhD thesis, Cracow University of Technology, Cracow (2013)
3. National Traffic Model elaboration by Warsaw University of Technology.commissioned by GDDKiA under project 'Studium układu dróg szybkiego ruchu w Polsce' (2008)
4. Rosik, P., Stępniak, M.: Monitoring of changes in road potential accessibility at municipality level in Poland, 1995-2015. Geogr. Pol. **88**(4), 607–620 (2015)
5. Rosik, P., Komornicki, T., Stępniak, M., Śleszyński, P., Goliszek, S., Pomianowski, W., Kowalczyk, K.: Evaluation of accessibility changes in Poland using the MAI indicator. Geogr. Pol. **90**(3), 361–368 (2017)

6. Rosik, P., Komornicki, T., Goliszek, S.: Traffic modeling in Poland at the municipal level. Multi-purpose model. In: Sierpiński, G. (ed.) Integration as Solution for Advanced Smart Urban Transport Systems. 15th Scientific and Technical Conference "Transport Systems. Theory & Practice 2018". Selected papers, Advances in Intelligent Systems and Computing, vol. 844, pp. 129–140. Springer, Cham (2018)
7. Rosik, P., Komornicki, T., Goliszek, S., Śleszyński, P., Szarata, A., Szejgiec-Kolenda, B., Pomianowski, W., Kowalczyk, K.: Kompleksowe modelowanie osobowego ruchu drogowego w Polsce. Uwarunkowania na poziomie gminnym, Prace Geograficzne, 267, Instytut Geografii i Przestrzennego Zagospodarowania PAN, Warszawa, 242 p (2018)
8. Żądło, M.: Wpływ gęstości podziału na rejony komunikacyjne na wyniki prognozy ruchu. Zeszyty Naukowo-Techniczne Stowarzyszenia Inżynierów i Techników Komunikacji w Krakowie. Seria: Materiały Konferencyjne. **1**(97), 245–256 (2012)
9. Stępniak, M., Rosik, P.: The impact of data aggregation on potential accessibility values. In: Geoinformatics for Intelligent Transportation. Lecture Notes in Geoinformation and Cartography, pp. 227–240. Springer (2014)
10. Rosik, P.: Dostępność lądowa przestrzeni Polski w wymiarze europejskim. Prace Geograficzne. **231**, 310 Warszawa (2012)
11. Więckowski, M., Michniak, D., Bednarek-Szczepańska, M., Chrenka, B., Ira, V., Komornicki, T., Rosik, P., Stępniak, M., Szekely, V., Śleszyński, P., Świątek, D., Wiśniewski, R.: Road accessibility to tourist destinations of the Polish-Slovak borderland: 2010-2030 prediction and planning. Geogr. Pol. **87**(1), 5–26 (2014)
12. Śleszyński, P.: A geomorphometric analysis of Poland on the basis of SRTM-3 data. Geogr. Pol. **85**(4), 47–61 (2012)
13. Śleszyński, P.: Expected traffic speed in Poland using Corine land cover, SRTM-3 and detailed population places data. J. Maps. **11**(2), 245–254 (2015)
14. Rosik, P., Stępniak, M., Wiśniewski, R.: Dojazdy do pracy do Warszawy i Białegostoku-alternatywne podejścia metodologiczne. Studia Regionalne i Lokalne. **2**(40), 77–98 (2010)
15. Głogowski, C., Zipser, T.: Formy i charakterystyki parametrów splotowego modelu pośrednich możliwości dla ruchów do pracy w dużych aglomeracjach. Zeszyty Naukowo-Techniczne Stowarzyszenia Inżynierów i. Techników Komunikacji w Krakowie. Seria: Materiały Konferencyjne. **1**(97), 77–100 (2012)
16. Okraszewska, R., Romanowska, A., Jamroz, K.: The effect of university campuses on the modal split of Polish cities. In: Intelligent Transport Systems and Travel Behaviour. Advances in Intelligent Systems and Computing, vol. 505, pp. 65–74. Springer, Heidelberg (2017)
17. Brzuchowska, J., Sławski, J.: Analiza rozkładów przestrzennych podróży "dom-szkoła" we Wrocławiu. Zeszyty Naukowo-Techniczne Stowarzyszenia Inżynierów i Techników Komunikacji w Krakowie. Seria: Materiały Konferencyjne. **1**(97), 5–20 (2012)
18. Michniak, D., Więckowski, M., Stępniak, M., Rosik, P.: The impact of selected planned motorways and expressways on the potential accessibility of the Polish-Slovak borderland with respect to tourism development. Moravian Geogr. Rep. **23**(1), 13–20 (2015)
19. Dudek, M.: Doświadczenia zagraniczne w modelowaniu ruchu związanego z wykorzystaniem czasu wolnego. Zeszyty Naukowo-Techniczne Stowarzyszenia Inżynierów i Techników Komunikacji w Krakowie. Seria: Materiały Konferencyjne. **1**(97), 21–40 (2012)
20. Szejgiec, B., Komornicki, T.: Spatial differentiation of Polish export linkages. Geogr. Pol. **88**(1), 173–178 (2015)
21. Marengo, G., Nilsson, S., Picco, P.: The "Territorialization" of mega transport infrastructure projects—the results achieved by the Turin–Lyon technical observatory. In: Fabbro, S. (ed.) Mega Transport Infrastructure Planning, pp. 203–214. Springer (2015)
22. Hancock, P.A., Nourbakhsh, I., Stewart, J.: On the future of transportation in an era of automated and autonomous vehicles. Proc. Natl. Acad. Sci. U. S. A. **116**(16), 7684–7691 (2018)

Relationship Between Transport Performance Indicators and GDP in Poland

Grzegorz Krawczyk

Abstract The purpose of the article is to analyse the relationship between transport performance indicators and GDP in Poland. The research takes into consideration the impact of changes in freight transport performance and traffic volume on GDP. Particular emphasis has been put on the analysis of correlation between GDP and road traffic volume. The main thesis of the article is that the data obtained from the measurement of road traffic in the micro scale can be used for predicting the GDP volume at the regional and national level.

Keywords Gross domestic product · Annual traffic · Freight transport performance · Road traffic volume · Cargo transport

1 Introduction

Transport system is strongly linked to the economic growth. The traces of this linkage appeared already in the first studies that comprehensively addressed the problems of transport economics. In the work entitled *Railway Economy* (1850), D. Lardner noticed the correlation between the volume of sales market and the level of transport costs, concluding that the sales volume of a particular good changes in a relationship which is inversely proportional to the square of the amount of transport rate for this good. Despite the criticism of the mathematical aspect of this statement [1], it shall be noted that the interconnectedness of economic matters and transport has already been the subject of research since mid-nineteenth century. In the work entitled *Verkehrsmittel*, another pioneer of transport economics, Emil Sax,

G. Krawczyk (✉)
University of Economics in Katowice, Katowice, Poland
e-mail: grzegorz.krawczyk@ue.katowice.pl

M. Suchanek (ed.), *Transport Development Challenges in the 21st Century*, Springer Proceedings in Business and Economics,
https://doi.org/10.1007/978-3-030-50010-8_4

presented the law of transport intensity, according to which the transport growth rate (particularly in the field of infrastructure) should be consistent with the development rate of the other sectors of economy [2]. Transport is defined as the lifeblood of economy and it is necessary for its proper development. It also constitutes an important element of shaping the quality of life for the inhabitants. This paper analyses the correlation between the development of transport and GDP in Poland. The development of transport system has been expressed by means of the indicators of transport performance and traffic volume in road transport. According to the thesis put forward in the paper, traffic volume is correlated with the GDP volume, whereas the results of traffic measurements can be used for predicting the economic condition of specific regions.

Traditionally, the literature concerning transport economics expresses the view that transport volume (expressed by means of different measures) is linked with Gross Domestic Product (GDP). However, it is difficult to determine the direction of this causality. In the research on Granger-casuality between transportation and GDP using a panel data set for the EU-15 states in the period 1970–2008, the conclusion was that only relatively well-developed economies demonstrate a bidirectional causality relationship between GDP and inland freight transportation per capita in ton-km, whereas a majority of the analysed countries have shown mixed results [3]. On the other hand, the research conducted in China demonstrated that bidirectional causality relationships between freight transport and economic growth had been detected in less developed central and western regions, whereas in the more wealthy and developed eastern region, it was demonstrated that economic growth was the reason for the growth of freight transport [4]. On the other hand, the research conducted for 27 countries with the so-called middle income from Europe, Asia and America demonstrated that the growth of GDP in 1994–2016 was responsible for a significant increase of road traffic volume and, consequently, for an increase in the number of road accidents [5]. The research on the causality has also been extended to other variables related to the transport system. The research conducted in the USA, in all counties of Minnesota, based on the data from the period 1988–2007, demonstrated that there was a bidirectional causality relationship between the population size and the local road traffic volume [6]. Therefore, the investments in transport network may also constitute a significant element in building strong economy and its ongoing development. The capability of infrastructural investments to generate economic growth depends on the general level of development of transport network. In developed countries, with expanded transport network of high quality, further investments in this infrastructure will not result in economic growth by themselves [7]. Undoubtedly, the correlation between transport and economy is not only limited to the relationships between transport performance and GDP. The research of National Travel Survey (NTS) conducted in the Great Britain has shown that the mobility of household members (measured with the number of journeys made per year) increases along with the growth of household income [8].

As can be noticed, the correlation between GDP and transport performance has been universally confirmed; however, the direction of such correlations in a causality is not known. The impact of transport on economy and economy on transport is varied due to the level of development of a particular country or region. Even in a group of relatively similarly developed EU-15 states, there is a noticeably different flexibility between the economic growth and the increase of transport performance measured as the ratio of the growth of transport performance and GDP growth [9]. The calculation of the value of transport performance is a relatively complex task carried out after an analysis, which entails certain delay in obtaining the result. In the further part of the paper, the question was asked to what extent the road traffic volume is related to GDP and to what extent such data can be used for estimating the economic condition of the region.

2 Correlation Between GDP and Transport Performance in Poland

Gross Domestic Product (GDP) is one of the basic measures of the effects of work performance of the society in a particular country. This indicator describes the aggregated value of ultimately manufactured goods and services in a specific country per unit of time. It is one of the most popular indicators describing the economic size of a particular country, whereas any changes of this indicator are used to determine the level of economic growth in a specific period of time. GDP is expressed in monetary values: in fixed prices (without inflation) and in current prices (with inflation); therefore, these values may differ from each other.

Transport performance, depending on the type of transport, is expressed as the product of two values: weight of transported cargo and distance at which such transport was executed, or the number of passengers carried and distance at which such transport was executed.

Table 1 presents the values of freight transport performance for Poland in 2001–2016, with special consideration of road freight transport. Both the values of freight transport performance and the volumes of GDP in current prices have shown an increasing trend in the analysed period. The share of road transport in total freight transport in Poland in the analysed period has significantly increased. This illustrates the changes in the division of modal freight transport which have occurred over the years, in consequence of which road transport has gained a dominant position in cargo handling.

Based on the above data, an analysis of correlations between GDP and transport performance in freight transport was performed (Fig. 1). For this purpose, the value of GDP was selected as the dependent variable, whereas the value of annual freight transport performance and the volume of annual freight transport performance in road transport were adopted as the explanatory variables. For the above-stated pairs of variables, the coefficient of determination (R^2) was determined for the linear model. This coefficient explains what percentage of dependable (dependent) variable is explained by means of independent (explanatory) variable. The results have been presented in diagrams nos. 1 and 2.

Table 1 Freight transport performance in road transport in Poland in relation to GDP

Year	GDP current prices (PLN)	Freight transport performance (mln ton-km)	Freight transport performance in road transport (mln ton-km)	Share of road transport in total freight transport performance (%)
2001	779,975	256,094	77,228	30.16
2002	810,617	254,324	80,318	31.58
2003	845,930	260,822	85,989	32.97
2004	933,062	290,704	110,481	38.00
2005	990,468	228,023	119,740	52.51
2006	1,069,824	248,711	136,490	54.88
2007	1,187,605	267,309	159,527	59.68
2008	1,286,069	279,172	174,223	62.41
2009	1,372,208	282,909	191,484	67.68
2010	1,445,298	308,073	214,204	69.53
2011	1,566,824	318,474	218,888	68.73
2012	1,629,425	325,775	233,310	71.62
2013	1,656,895	347,887	259,708	74.65
2014	1,720,430	348,022	262,860	75.53
2015	1,800,228	360,635	273,107	75.73

Source: Data of Statistics Poland

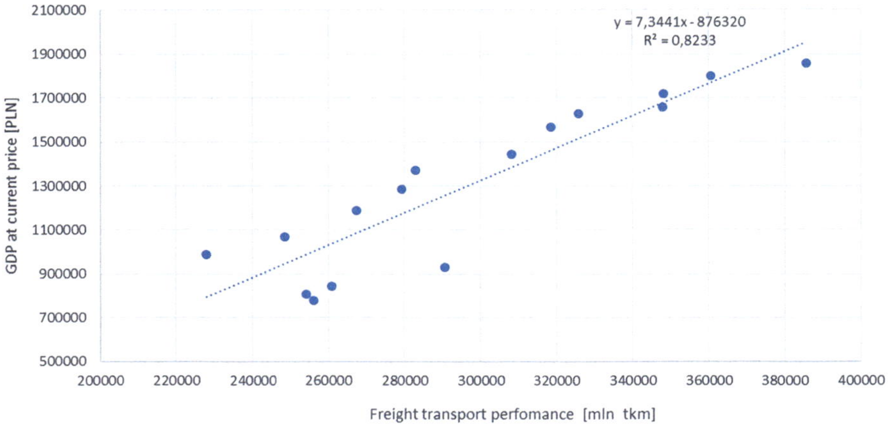

Fig. 1 Relationship between freight transport performance and GDP in Poland

Based on the above analyses, it may be concluded that the variability of GDP in Poland for 2001–2016 was confirmed in 82.33% with the variability of the volume of freight transport performance and in 99.17% with the variability of the volume of freight transport performance in road freight transport (Fig. 2). The statistical correlation between GDP and freight transport performance is very high in the analysed period, especially in case of road transport.

The positive correlation between freight transport performance and GDP has also been indicated in other EU states. The total analysis for 28 EU states in 2009–2015 demonstrated that the value R^2 between GDP in fixed prices and freight transport performance amounted to 0.5356, whereby this indicator for the so-called new EU states (the countries which have joined EU since 2004) amounted to 0.9108, while for the so-called "old" EU member states (the first 15 member states), the value of R^2 was 0.6132 [10].

3 Correlation Between GDP and Traffic Volume in Poland

This part of the paper presents the analysis of the correlation between the volume of GDP in particular regions of Poland and the level of road transport volume. It is difficult to determine the volume of transport performance which is the product of either distance and number of passengers, or distance and cargo weight. The fundamental difficulty is the collection of precise data about the volume of transport and the time needed for their analysis. Transport companies, with certain exceptions, are not obliged to report to public administration units any data concerning the length of journeys made, number of passengers carried, tons of cargo, or distances of completed routes. In consequence, the analyses concerning the volume of transport performance are based on specific models relying on a certain narrow scope of economic data. In reality, it is much easier and quicker to define the traffic volume as such, which should also illustrate the broadly understood state of economy. In order to verify this hypothesis, the correlation between the volume of GDP for the region and the data obtained from the measurement of traffic volume was analysed.

The data concerning GDP at the regional level were presented for specific provinces. Since 1999, there has been a three-level administrative division of Poland into provinces, powiats (counties) and gminas (communes or municipalities). Thus, province is the administrative unit of the highest order. According to the classification of territorial units for statistical purposes, province represents the level NUTS 2. Poland is divided into 16 provinces, which were formed in line with the historical and geographical conditions. The provinces differ from each other in many aspects; in particular, there is a division into richer areas in western Poland and poorer eastern part of the country. The differences, especially in respect of the urbanisation level and infrastructural equipment, mainly result from the history of these regions [11]. Table 2 presents selected parameters that characterise the provinces in Poland.

The territorial units described in the table differ from each other with multiple indicators, including population density and transport network density. There are also differences indicated by the comparison of the degree of urbanisation, which determines the percentage of inhabitants registered in the cities. The above-quoted short characteristics indicate that the provinces in Poland differ from each other also in terms of GDP and traffic volume.

The main institution which deals with the problem of traffic measurement in a complex and nationwide manner is the General Directorate for National Roads and

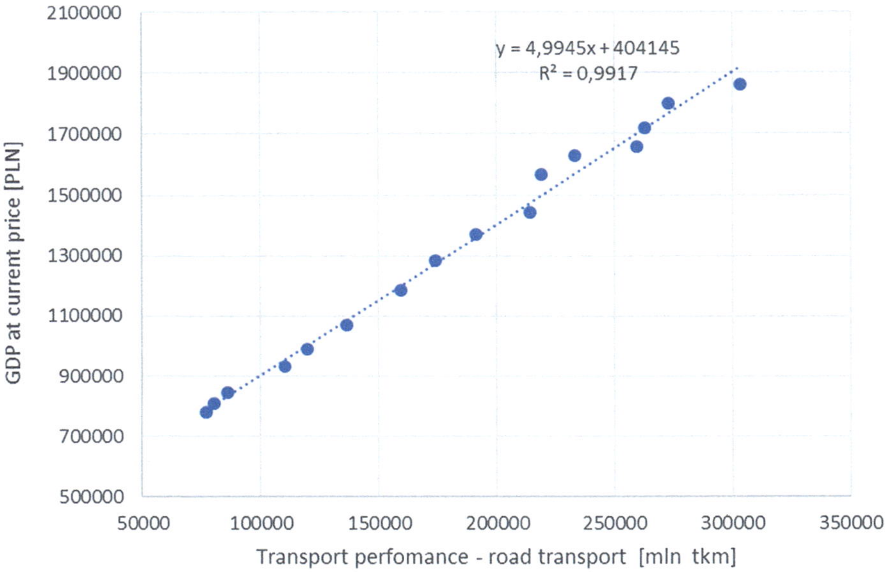

Fig. 2 Relationship between road freight transport performance and GDP in Poland

Motorways (GDDKiA). This entity is an office of government administration whose tasks include e.g. implementation of transport development policy in respect of road network, traffic management on national roads, as well as supervision and coordination of infrastructural investments in respect of national roads and motorways. GDDKiA manages continuous traffic measurement stations. There are currently 103 measurement stations in which data are collected continuously throughout the year. The data are verified and analysed on an ongoing basis, and a study is prepared once a year in order to summarise the performance of the whole system and present its results. The value of annual average daily traffic (AADT) is determined on the basis of the performed measurements. AADT values calculated on the basis of data from the continuous traffic measurement stations are characterised by high accuracy. AADT value takes into consideration all vehicles, regardless of their type and size. It has been decided that the values obtained in this manner, averaged for the entire calendar year, will be compared against the GDP values obtained in specific provinces.

In order to present the relationships between traffic volume and GDP as precisely as possible, the decision has been made to identify AADT values in 2006–2016, which overlaps with the availability of data concerning the level of GDP in specific provinces. Despite the developed system of continuous traffic measurement stations, the main problem is to obtain data from one location over a longer period of time. The system of stations has gradually evolved, whereas in some of the provinces continuous measurement has only been performed for a couple of years. The stations were located within the administrative borders of the largest urban centres

Table 2 Characteristics of provinces in Poland (status as of 31 December 2018)

Province	Population density (person/km²)	Urbanisation level (%)	Public road density[a] (km/100 km²)
Lower Silesia	145.45	68.61	98.5
Kuyavian-Pomeranian	115.62	59.09	102.1
Lublin	84.29	46.46	88.7
Lubusz	72.53	64.95	63.3
Łódź	135.37	62.55	113.7
Lesser Poland	223.98	48.19	170.6
Mazovia	151.96	64.40	106.8
Opole	104.82	53.30	90.3
Subcarpathian	119.30	41.09	94.2
Podlaskie	58.53	60.79	66.2
Pomerania	127.44	63.66	75.9
Silesia	367.59	76.73	176.8
Świętokrzyskie	106.02	44.86	123.4
Warmian-Masurian	59.11	58.99	56.4
Greater Poland	117.14	54.27	98.8
West Pomerania	74.31	68.50	61.2

Source: Data of Statistics Poland
[a]Data concerning the road density for 2017

in a specific province. Currently, on the average, there are several continuous traffic measurement stations in each province; unfortunately, due to the above-described conditions, the results from one station per province only were chosen for further analysis. The choice of the station was based on the criterion of availability of measurement results for each year in the analysed period; the measurements carried out on the motorways were not taken into account. Consequently, the data from eight locations were selected for further analysis (in line with Table 3). For the other provinces, the data were incomplete and not used for further analysis.

The measurement stations from which the results will be subject to further analysis are located in big cities—the province capitals. Obviously, these results do not directly reflect the economic condition of a particular area. The volume of traffic flows expressed with AADT indicator is influenced, e.g. by class and technical parameters of the road, its significance in the local and regional system, as well as transit traffic. Therefore, it cannot be confirmed that the volume of traffic flow measured at one point reflects the economic situation of the entire province. There is no correlation between the average AADT values presented in the table and the average GDP value in 2006–2016. However, the most important point of this argument is to determine to what extent the variability of traffic flows is correlated with the variability of GDP volume. The division and characteristics of the variables used in the study were presented in Table 4.

For the purpose of the calculations, the values describing traffic volume were adopted as explanatory variables, whereas the values related to GDP were adopted

Table 3 Location of measurement points and average AADT value

Province	City	Road	Average AADT for the period 2006–2016 (vehicle/day)
Lesser Poland	Kraków	DK 52	14,404
Mazovia	Warsaw	DK 2	11,939
Podlaskie	Białystok	DK 61	9742
Pomerania	Gdańsk	DK 20	15,743
Świętokrzyskie	Kielce	S 71	17,358
Warmian-Masurian	Olsztyn	DK 16	8910
Greater Poland	Poznań	DK 92	9853
West Pomerania	Szczecin	DK 6	13,885

Source: Data of Statistics Poland and General Directorate for National Roads and Motorways

Table 4 Variables adopted for the analysis

Variable	Description
X_1	Annual AADT volume in the measurement point
X_2	Change of AADT volume (previous year = 100%)
Y_1	GDP of the province, annual, in current prices
Y_2	GDP of the province, per capita, annual, in current prices
Y_3	Change in GDP volume of the province, per capita, in current prices, previous year = 100%
Y_4	Change in GDP volume of the province, in fixed prices, previous year = 100%
Y_5	Change in GDP volume of the province, per capita, in fixed prices, previous year = 100%

Source: own study

as dependent variables. The following indicators were determined for these values: Pearson correlation coefficient (P) and coefficient of determination (R^2) for the linear correlation.

Based on the calculations, a relatively high correlation was noticed between AADT and GDP values in current prices, both for the province and per capita. The results were presented in Table 5. The correlation between the remaining pairs of variables did not demonstrate a significant statistical correlation (in a majority of cases the value of coefficient of determination did not exceed 0.3).

In a majority of the analysed cases, the values of correlation and determination coefficients are high. The lowest correlation between the road traffic volume and GDP was reported in Lesser Poland and Pomerania Provinces. In the other analysed provinces, the relationships were stronger. Obviously, an analysis of this type should take into consideration changes in the road system. Construction of a new road or temporary changes in traffic and redirection of additional traffic flows to a section covered by measurement will result in obtaining incorrect analysis results. In a majority of the analysed cases, the correlation between the traffic flow volume determined by AADT indicator and GDP in current prices was high. The weak point

Table 5 Correlation between selected variables

Province	Variable	Value of correlation indicators with variable X_1	
		P	R^2
Lesser Poland	Y_1	0.711	0.506
	Y_2	0.709	0.503
Mazovia	Y_1	0.876	0.767
	Y_2	0.872	0.762
Podlaskie	Y_1	0.919	0.845
	Y_2	0.918	0.842
Pomerania	Y_1	0.660	0.436
	Y_2	0.653	0.427
Świętokrzyskie	Y_1	0.908	0.824
	Y_2	0.894	0.799
Warmian-Masurian	Y_1	0.911	0.830
	Y_2	0.917	0.840
Greater Poland	Y_1	0.985	0.970
	Y_2	0.986	0.972
West Pomerania	Y_1	0.921	0.848
	Y_2	0.922	0.850

Source: own study

of the analysed issue is a relatively small amount of data. It was difficult to obtain the measurements of traffic volume from the same measurement point over a long period of time. Due to this, some of the provinces were not included in the analysis, whereas the results used in the calculations were more sensitive to deviations. The quality of calculations and conclusions drawn on their basis will increase along with the growth of the network of measurement points.

4 Summary

Based on the presented calculations and conducted literature review, it may be concluded that there is a correlation between the traffic flow and GDP value. On the basis of the presented calculations, it is impossible to form a conclusion regarding the direction of causality; this problem constitutes another interesting research area. The arguments in favour of using traffic measurements for predicting the economic results at the regional level include speed and accuracy of road traffic volume (in case of having an expanded network of measurement points), possibility to obtain results at the regional level, as well as obtaining results with a frequency higher than once a year.

References

1. Shieh, Y.N., Goldberg, I.: Lardner's law of squares. Econ. New Ser. **52**(208), 509–512 (1985)
2. Pickhardt, M.: Perspectives on Competition in Transportation, pp. 222–225. LIT Verlag, Munster (2006)
3. Beyzaltar, M.A., Karacal, M., Yetkiner, H.: Granger-casuality between transportation and GDP: a panel data approach. Transp. Res. A Pol. Pract. **63**, 43–55 (2014)
4. Tong, T., Yu, T.E.: Transportation and economic growth in China: a heterogeneous panel cointegration and causality analysis. J. Transp. Geogr. **73**, 120–130 (2018)
5. Qamar, A., Yaseen, M., Khan, M.: The casuality of road traffic fatalities with its determinants in upper middle income countries: a continent-wide comparison. Transp. Res. A Pol. Pract. **119**, 301–312 (2019)
6. Iacono, M., Levinson, D.: Mutual causality in road network growth and economic development. Transp. Pol. **45**, 209–217 (2016)
7. Banister, D., Berechman, Y.: Transport investment and promotion of economic growth. J. Transp. Geogr. **9**(3), 209–218 (2001)
8. Metz, D.: Saturation of demand for daily travel. Transp. Rev. **30**(5), 659–674 (2010)
9. Tapio, P.: Towards a theory of decoupling: degrees of decoupling in the EU and the case of road traffic in Finland between 1970 and 2001. Transp. Pol. **12**(2), 137–151 (2005)
10. Gnap, J., Konecny, V., Varjan, P.: Research on relationship between freight transport performance and GDP in Slovakia and EU states. Nase More. **65**(1), 32–39 (2018)
11. Yoder, J.A.: Decentralisation and regionalisation after communism: administrative and territorial reform in Poland and the Czech Republic. Europe-Asia Stud. **55**(2), 263–286 (2003)

Value Creation for Stakeholders in the Implementation of "Last Mile" Deliveries in Cities

Blanka Tundys and Piotr Niedzielski

Abstract Developing of e-commerce services market and increasing volume of parcels is one of the biggest challenges in the field of order execution and "last mile" logistics. Ensuring the highest quality of services and comfort for the customer of the B2C market has an impact not only on the quality of the entire logistics process, but also on the image of service providers. Consumers are increasingly paying attention not only to product price, time, and cost of delivery. An increasingly important role plays the aesthetics of the shipment itself, the declared date of delivery, completeness of the package, driver's behavior, flexibility of delivery, or ease of picking up the ordered goods. The aim of the discussion is to indicate how to create value for stakeholders and supply chain participants by implementing a new business model supporting the "last mile" delivery of goods in the cities. The research was based on the project of an "open" network management system "delivery box" allowing access to the infrastructure for many service providers. The results of the conducted pilot studies allow to distinguish the preferences of customers in terms of the service of the "last mile", but also possible problems and barriers to the implementation of a new business concept.

Keywords Last mile · Value creation · City delivery · Management · Stakeholders · Delivery box

B. Tundys (✉) · P. Niedzielski
Faculty of Economics, Finance and Management, University of Szczecin, Szczecin, Poland
e-mail: blanka.tundys@usz.edu.pl; piotr.niedzielski@usz.edu.pl

M. Suchanek (ed.), *Transport Development Challenges in the 21st Century*, Springer Proceedings in Business and Economics,
https://doi.org/10.1007/978-3-030-50010-8_5

1 Introduction

The "last mile" problems are one of the elements of the implementation of the city logistics concept and the smart city idea. Along with the growing number of inhabitants in cities, the problem of supplying consumers with goods is growing, especially taking into account the service of individual customers, who buy through the e-commerce market. In the context of the development of this market, the service of the last mile will be increasingly important from the point of view of the efficiency of processes occurring in the whole supply chain. The structure of modern cities, including limited availability of parking lots and congestion are unfavourable for suppliers, increasing their costs as a consequence [1]. This requires to search for new, effective and innovative solutions.

The increased interest in the e-commerce market contributed to the increase in demand for last mile deliveries, and the creation of a specific model where the customer either receives a parcel at home or from a collection point (parcel lockers) causes not only a change in the market and the introduction of a new service, but also other ways of creating value for all stakeholders (operators, customers, the general public). Such solutions indicate that effective management of the last mile is crucial for stakeholders, as this stage accounts for approximately 7% of the total amount of financial resources involved in parcel delivery, and these are primarily costs incurred by carriers [2]. The number of vehicles transporting goods is growing rapidly in the cities. And competition on the e-commerce market contributes to shortening the delivery time in which the order was placed (same day delivery). This element leads to destabilization of the traditional way of functioning of the logistics sector, change of the perception of industrial real estate, and transformation of "last mile" solutions. In 2018, 55% of the human population, or 4.2 billion people, lived in urban areas. UN forecasts indicate that by 2050 it will have grown by 2.3 billion more. This means that the proportion of the urban population will rise to 68% of the total population [3]. Cities of the European Union are inhabited by 75% of the population, and in Poland by about 60% [4]. The population is concentrated around large cities. This determines the development of the logistics sector and its implementation of innovative solutions. City dwellers are usually active and demanding people, often reaching for non-standard innovative solutions when making purchasing decisions. This means that the organization of supply chains is under increasing pressure to deliver goods earlier and faster to meet the ever-changing consumer demand. The development of the e-commerce market has contributed to the complexity of supply chains. Inventory rotation cycles have been significantly shortened and at the same time more unpredictable. The new model of market functioning, with particular emphasis on purchasing on the e-commerce market, contributed to changes in customer expectations and the need to adjust the supply chain strategy to them. What becomes important is not only the availability of goods, but also the necessity to strive for operational excellence of the chain, including the so-called "last mile" deliveries. The challenge for chains is the need to be close to the market or to be able to carry out tasks properly in order for the satisfaction of the

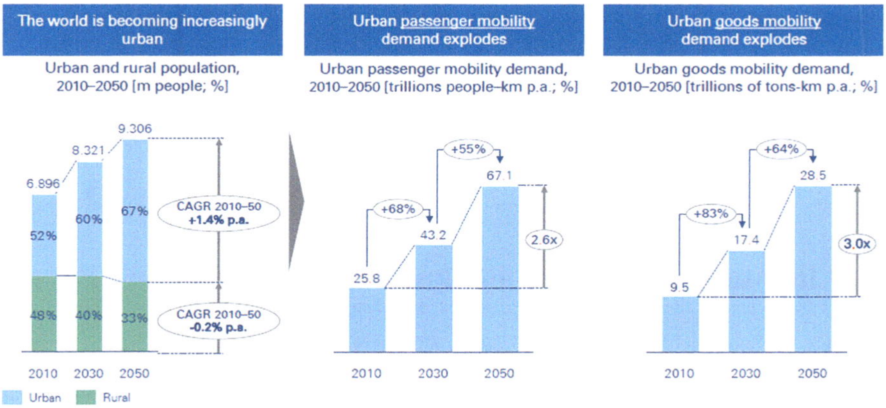

Fig. 1 Increased urbanization and its impact on passenger and goods mobility demand. (Source: UN, World Bank, OECD, ITF, Schäfer/Victor 2000, Cosgrove/Cargett 2007, Schäfer 2007, Arthur D. Little)

final consumer. Supplies coming from the e-commerce market and taking place on the so-called "last mile" must create added value for stakeholders. These are not only consumers, but taking into account the specificity of the chain, they come to the forefront of the recipients of the value of the new market. Figure 1 shows how both the urban population and the demand for mobility of passengers and freight are changing. Increasing trends are clearly visible (both in terms of population growth and demand for freight and human mobility) and projections indicate that by 2050, compared to 2010, urban passenger mobility will increase by 2.6 times and freight by 3 times. Although Fig. 1 presents only forecasts, this type of trends can definitely be observed, especially in the context of the constantly increasing volume of cargo, which have its destination in the city and the expansion of the e-commerce market. Taking into account the changing demand of customers, the expectation of innovation and added value while not increasing costs, it is necessary to look for new business solutions that will bring added value to all stakeholders.

The problems described above have contributed to the creation of the following part of deliberations: (a) characteristics of theoretical aspects related both to the theory of stakeholders and creating value for them, as well as the connection of a new supply chain strategy with these activities, the focus of which will be on creating value within the "last mile" delivery with a focus on servicing e-commerce; (b) analysis and interpretations of pilot studies conducted by authors, related to the use of new infrastructure and a new business idea for servicing the "last mile" delivery in cities; (c) indication of the possibility of creating value for stakeholders, as well as limitations. The authors adopted the methodology of literature analysis and pilot empirical studies, which were carried out with the use of a questionnaire. Next, conclusions were drawn using descriptive statistics methods. The considerations are accompanied by the following research questions: whether it is possible to identify ways of creating added value for stakeholders within the "last mile" deliveries in cities and whether new business models based on the extension of functionality and availability of already existing solutions for parcel delivery in cities will contribute to increasing not only the value for customers but also for other participants in the supply chain, and thus will contribute to increasing the competitiveness of new solutions, increasing the efficiency of chains and reduce the negative impact, e.g. on the environment, "last mile" deliveries in cities.

2 Literature Background and Problem Description

The dynamic development of information technologies contributed to the development of the e-commerce market. The e-commerce revolution is transforming business processes, supplier and procurement systems, the structure of industries, and the nature of firms [5]. In Poland, the dynamics of sales in e-commerce is constantly increasing, maintaining the double-digit dynamics of this growth, and at the same time constituting the fifth e-commerce market in Europe after the UK, Germany, France and Italy [6] such solutions influence the design of supply chains and the adaptation of technology that enables the integration of partners' activities, greater transparency, sharing of information and changing cooperation models [7]. In the context of the creation of new markets, innovative business models are also being developed which aim, among other things, to create added value for all stakeholders. They are supported by specific supply chains operating on the e-commerce market. They differ from the traditional supply these, that in cooperation with the e-commerce platform products can be directly sent to the manufacturers, not being the property of the e-commerce platform, which is only an intermediary and provides sales service and receives commission income [8]. Consumers also pay more attention to the handling and service that affects the market demand for the products concerned [9], as well as e-commerce, it changes the number of enterprises that are links in the supply chain, highlighting the need to meet consumer demands [10]. E-commerce supply chains must solve the problems of large volumes, often small supplies, to densely populated urban areas. Acting in such a specific supply chain is not an easy task to create added value, especially in the context of solving the "last mile" problem. New business model is a combination of the value stream, revenue stream and logistical stream of a business in which the value component drives the revenue and logistical components of the firm [11]. It is necessary to connect the issues connected with the supply chain management connected with e-commerce and solving the "last mile" problems. Online shopping allows for convenience, simplicity, better access to information and time efficiency [12]. In this context, companies in supply chains also benefit from the absence of geographic boundaries, optimization of information and flow of goods, lower costs of supply, transactions and advertising [13]. Implementing a new business model and in order to reduce the costs of delivery of ordered parcels to home, logistics operators use other alternative solutions. One of them is to create intelligent packet lockers in convenient locations (at universities, schools, shops, office buildings, large housing estates) [14].

Operating in the supply chain is not an easy task, especially in the context of solving the "last mile" problem. Therefore, creating business models based on fast, reliable deliveries, often in the same day delivery formula and using modern equipment for collecting and sending parcels, with IT support, requires creating a new value for all participants. The growth of retail trade has contributed to an increase in the volume of deliveries, but also in returns, which at the same time puts pressure on the last mile's operators. In business and e-commerce, customers are heavily involved in value creation [15]. Each company, in pursuit of a universal goal, i.e. to continue and develop on the market, carries out value-added processes which are directly or indirectly related to the production of goods or services and thus creates a specific value for different groups of stakeholders (e.g. customers, employees, owners, suppliers). At the same time, stakeholders perceive the value provided by the company in different ways. This aspect should be highlighted and added value should be sought by creating new business models. Understanding by the company, or in a broader context, by the supply chain, the essence of the created value for different groups of stakeholders, and not only for one group such as customers, its sources and mechanisms of formation is a condition not only for effective management but also to create competitiveness of the company.

Implementation in value creation processes of resources being outside the company's borders, including other stakeholders, is increasingly used by companies in the strategy of creating competitiveness on the market [16]. The ability to use resources by an enterprise, being outside the boundaries of the organization, is becoming more and more important as a determinant of shaping the competitiveness of enterprises. The competition is based on: providing additional services by attaching to the basic offer accompanying services, attractive prices—by differentiating their amount, forms and conditions of payment, appropriate distribution—by improving forms, distribution channels and customer service, including after-sales service, communication—by providing fast and reliable information. In order to survive and develop on the market, an enterprise must cooperate with various stakeholder groups, the scope of cooperation depends to a large extent on the maturity of the parties to the cooperation, but the motor of mutual cooperation is the willingness to obtain the synergy effect [17]. By working with stakeholders, a company creates value/proposals for value, which is a combination of a number of factors that include: price offers of products/services offered to customers by the company, the quality of relations with individual stakeholder groups, the company's image, as well as new solutions and business models offered to stakeholders. The starting point for managing the stakeholder value creation process is the proper management of stakeholders and their relations with the partners. The stakeholder management process is a continuous process and consists in adapting to new threats and opportunities on the part of stakeholders and changing the strategy towards existing entities. It is also important that in new business models, especially when considering relations in the supply chain, one should not only look for value at each stage, but also involve stakeholders at each stage, because only then can one achieve a competitive advantage and effectively develop innovative business models. Therefore, regardless of the size and role of the partners, their knowledge, skills and the possibility of implementing new solutions contribute to the value creation.

Courier services provided in cities can be provided in several business models. Taking into account the e-commerce market service, the most important forms of service include: courier service of small parcels of relatively high value, constituting a subgroup of express services of low weight. One can speak of express services, delivered quickly and reliably, where the service includes strict delivery times, parcel services tailored to individual needs of customers, using standard service processes. Current business models provide a wide range of parcel delivery services. The main factors shaping business strategies and service models include [18]:

- Delivery time of the consignment to the recipient: with immediate delivery (immediate), the same day (same day), on the next working day (next day), on the second consecutive working day, as a rule in relation to deliveries of parcels and express shipments to selected countries abroad, (second day), with night delivery, at a guaranteed time and price of deliveries, collected in the evening and delivered no later than 10:00–12:00, the next day (overnight), with deliveries during night hours (at night), e.g. in the delivery of spare parts.
- Range of the operator: local services, regional services, domestic services, European-wide services, global services, intercontinental services.
- Business areas served: services diversified due to suppliers and recipients of supplies, indirectly influencing the business areas served, include: B2B services (Business-to-Business). They are the domain of express companies, e.g. in Poland the vast majority of express mail is transported between shippers and institutional customers (production, trade or service companies); B2C services (Business to Customers)—concern deliveries of mail sent by institutional partners (including many small companies) to their end customers (consumers or users).
- Modes of transport used: express deliveries are most often made by road and air, where road transport usually handles local, regional and national deliveries, while air transport—deliveries in international and intercontinental relations.

The types of services provided within "the last mile" delivery are subject to continuous expansion, including the following: delivery of a shipment from the sender to the receiver or to a place designated by him, implementation by a specialized company, use of an appropriate operational (logistic) network, possibility of tracking information on the status of the shipment during its movement, a specified delivery time, delivery in the shortest possible time, maintaining appropriate costs of delivery, implementation of tasks in accordance with customer expectations [19]. Value-added and build-to-suit services (BTS) [20] currently constitute a wide range of additional services offered by entities providing such services. The elements that create added value for stakeholders, in this case the ultimate customers, may include: additional insurance; collection and delivery on Saturday, Sunday and holidays at different times; additional attempts to deliver; notification of collection, delivery, postage; proof of delivery; preparation of the consignment for transport, including the provision by couriers of appropriate packaging for transport, e.g. the following Pharma Packs of goods that must be transported at a constant temperature (Thermo Boxes) and Dry Ice; services reducing emissions, CO_2; transport of shipments of non-standard shape and dimensions; transport of dangerous, heavy and oversized goods; transport to remote, not covered by the network of land deliveries; storage, service including storage of the consignment at the courier's premises; customs service; returns of documents attached to the original consignment; re-addressing during the delivery process; sending or receiving of conductor's consignments; telephone preference of the consignment by the courier and other services. More and more often in the context of last mile service parcel lockers are used (also for creating new values), which allow for self-service collection and return of goods purchased online [21]. Parcel lockers have received positive feedback from both consumers and businesses, improving the service experience for the former and providing competitive advantage and performance enhancement for the latter [22]. Self-service box parcel delivery services are expanding into national and international markets [23] and the growing interest may indicate a takeover of part of the deliveries in the future within the last mile [24]. The solutions for self-service transmitter/receiver boxes are scalable, customisable, electronic, and often cloud-based systems that give onsite and remote workers and users an easily accessible space for the retrieval of letters and parcels [25]. Therefore, it can be stated that the solution related to the delivery and collection of parcel from the box (parcel lockers) is the latest solution and trend I is a key tool in the context of handling the last mile in the cities for deliveries in e-commerce.

The structure of the KEP industry in Poland includes the following types of entities: KEP integrators, local KEP operators, postal operators, local operators, intermediaries. Currently, the concept of delivery of ordered goods, especially for individual customers, is much broader and includes delivery either of a specific place (e.g. a shop or a house) or a device (e.g. a parcel machine). Package sending and receiving points can become both an element that adds value for stakeholders

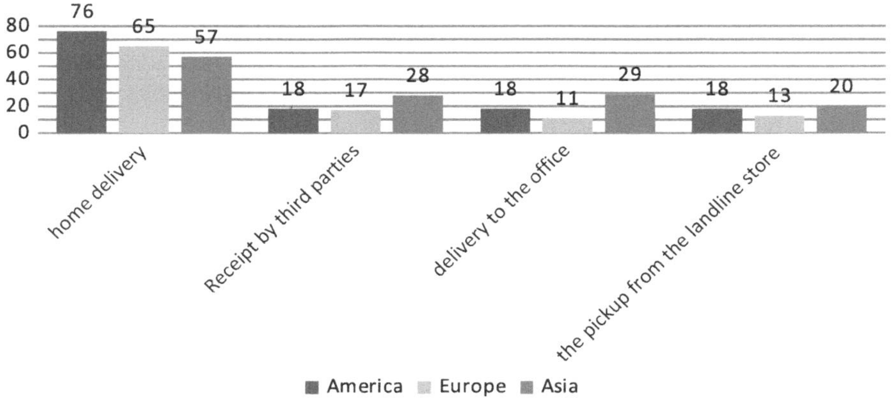

Fig. 2 The delivery method. (Source: [26])

and an element of competitiveness of a given entity. The forms of parcel collection were change. Only a few years ago, customers have paid attention that the package was delivered by a courier to the house. This form of delivery has been chosen most often so far (Fig. 2), but other available forms of pick-up or delivery are becoming more and more popular. The forms of location of pick-up points must be highly accessible, so they can be placed in shops, petrol stations and kiosks. The development of the B2C market and the increase in the popularity of e-commerce allowed to increase the number of such points. Self-service transmitting and receiving machines are also becoming more and more popular. Important selection criteria are: convenience and availability, but also low price and high quality of services, e.g. punctuality. The specificity of the market of parcels purchased on the e-commerce market by individual recipients, especially in large cities, is the small size of parcels and the lack of attachment to a single courier company, which allows for further development of new services and business models. New sales channels are being implemented, including the increasingly popular use of the Internet, within which the click & collect service is becoming increasingly popular, which allows you to place an order online and pick up the package in a selected location or device. The increase in the share of the online channel in retail sales and new supply models translate into a change in the type of shipments delivered. This includes another trend observed on the market, namely the need for same day delivery itself, which supports daily shopping by offering express deliveries, i.e. the same day. The same day delivery model itself must be based on an efficient and reliable pick-up system that can be offered by both traditional couriers and parcel collectors. Research on the use of lockers to collect parcels purchased on the e-commerce market is conducted in various contexts [27]. Solutions of this type are self-service tools that change the roles of the customer. They are both customers and creators of services. The customer is responsible for co-creating value for the customer [28]. This type of service allows customers to gain a lot of benefits, which can include: flexibility, unrestricted access (24 h a day), fast delivery, avoid time pressure to pick up the package from the courier and save time by not having to wait for the supplier and anonymity when picking up the package [8]. Considering the issues related to the use of inboxes, it can be indicated that their use can bring measurable benefits to three types of stakeholders. From the point of view of logistics operators and couriers, the use of intelligent self-service solutions eliminates inefficiencies caused by unsuccessful delivery (absence of a customer at home) and the need for re-delivery. In addition, the use of such solutions enables consolidation of shipments to clustered locations (i.e. self-collection hubs), and such solutions may improve the use of fixed assets, vehicles and reduce the number of journeys, which will not only contribute to improving the quality of air in cities, but also reduce traffic. Customers, thanks to the possibilities offered by the intelligent parcel lockers, avoid having to wait at home for shipment and delivery [29]. Customer shipments are temporarily stored in lockers and can be picked up at a convenient time. This allows you to reduce unnecessary alternative costs for customers (e.g. waiting time), which they can use in other ways. From the point of view of society as a whole, such solutions contribute to minimizing externalities such as already identified congestion,

congestion, noise and environmental pollution in general due to greater consolidation of shipments and fewer journeys by delivery vehicles in cities [12, 30].

"The Last Mile" is the final, extremely important and challenging part of the supply chain. The processes taking place in this area influence the evaluation of all the tasks performed (delivery of goods) in the understanding of the customer. It is particularly important in urban space, being the last link and "contact" of the supplier (courier, logistics operator) with the customers. The requirements set by consumers are driven by the e-commerce sector. Such a situation may be the most serious factor destabilizing the traditional functioning of the logistics sector through the reorganization of the supply chain, a complete change in the perception of industrial real estate (including city terminals) and the transformation of urban logistics solutions (such as the development of "Delivery Box-parcel locker"). Additionally, the "last mile" service is connected with the increasing number of replay information and re-delivery of parcels for the residents. Preferred delivery methods are presented in Fig. 3. The most preferred is home delivery, and further developing the home delivery system, also to use an automatic pick-up point, which may be located near the home.

Big data sets and the opportunities they offer in the context of the so-called "last mile" challenge are a source of creative solutions, such as tracking, sending, notifying and sending messages to drivers and consumers concerning the product, as well as providing delivery services to omni-channel sellers and self-receipt by the final customer (Fig. 3).

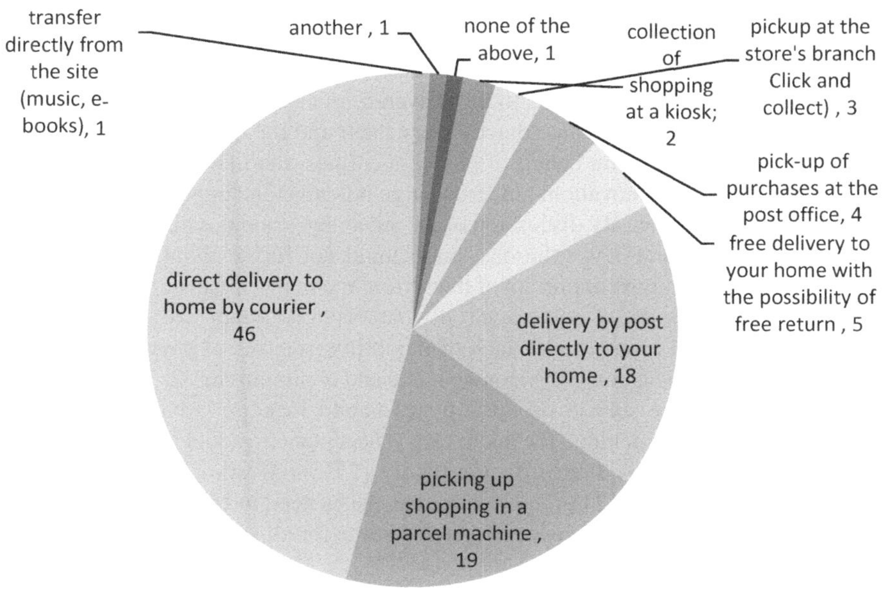

Fig. 3 Means of receipt of consignments (%). (Source: [4])

Innovative strategies developed for the solution of "last mile" delivery and used in densely built-up industrial centres around the world include multi-level warehouses, parcel/collection points and transhipment centres before final delivery to the recipient. Parcel boxes also have the potential to reduce environmental impact and increase the overall speed of service delivery [31]. Therefore, identifying and creating value for all stakeholders needs to include different services and expanding the range of services. One of such possibilities is to create a generally accessible delivery box'es system, which is a device for distribution of products of various types, including special products requiring e.g. low or high temperatures, various sizes and materials contained in parcels, operated by means of smartphone applications. The characteristic feature is the exact, estimated date and time of delivery of the parcel and the fact that it is available to every market participant (i.e. every courier who wants to use this service, or the company offering parcel delivery, as well as the individual customer). The implementation of the solution contributes to the possibility of reducing the environmental burden in the urban space. This is a result of rationalization of courier services in urban areas developed by courier companies. Thanks to the development and implementation of a project based on the concept of cooperation and sharing of resources by the main stakeholders of the distribution/supply system of courier parcels at the so-called "last mile" delivery in the city, will be implement new business model. In the context of a city, the concept can contribute to reducing the number of vehicles on urban roads, as an alternative form of supply for individual consumers. It contributes to more efficient deliveries, fewer advised parcels and the need to return to the indicated address, thus reducing the number of vehicles on the roads. It gives the customer great flexibility in picking up the parcel and, as a result, making an independent decision on where the parcel is to be delivered and at what times, the customer can or wants to pick it up. The system and delivery preferences depend on many factors, including the customer's wealth, but also on character of the goods to be delivered. In connection with the growing trend in the development of same day delivery itself and the limited possibilities of couriers already working on behalf of business entities, the indicated solutions are characterized by high innovation, but also a large potential and opportunities to exist and fill in the market gap. By discussion of the advantages and possibilities of developing new services and new solutions, one should not forget about the disadvantages they bring. The most important of them from the point of view of the customer include a limited possibility of storage (only parcels of a certain size), the possibility of vandalism and crime (theft), lack of flexibility in terms of payment [8], also human inertia, technical anxiety, inconvenience and exploitation [32, 33]. Scientific research related to the subject of smart boxes (smart lockers) is based on known theories, which may include: the technology acceptance model [34], theory of planned behaviour, innovation diffusion theory [25], motivation theory [35], and theory of reasoned action [21]. This means that issues need to be researched, developed and promoted so that the expected added value for all stakeholders can gradually increase.

3 Empirical Research

Within the framework of the SmarLab project, one of the objectives of which is to create and implement a system of open and generally available solution (delivery box) for sending and receiving parcels and solving the problem of the last mile. The authors carried out pilot studies, the purpose and task of which was to indicate whether this type of solution is useful on the Polish market of courier services and what should be the most important elements of such a business model. Another important element of the considerations and pilot studies undertaken was to indicate whether such a solution could bring value for stakeholders and, if so, for which groups.

The research was conducted, in period from October 2018 to June 2019, in the West Pomeranian and Pomeranian Voivodships, on a group of 169 respondents. During the survey, respondents were asked, among others, about the elements of added value. The most important of them for respondents were: accessibility and ease of use of the system, including the possibility of further online system and mobile application, convenience, reduction, costs, including time (both the execution of the order and waiting for delivery), adaptation to individual needs (including the possibility to choose the place and time of delivery), guarantee of faster delivery, facilitated access for all bidders, short delivery time, punctuality, reduction of service and service costs, convenience, flexibility, lower service price, solving environmental problems, comfort of reception, quality and speed of delivery, reliability, higher quality, ensuring customer satisfaction, time availability, ease of use, quick reception and dispatch. The results of the research respond positively to research questions, while being consistent with both theoretical foundations and management theories. This means that the presented theses in the confrontation of theory and practice are correct and that there is a need to undertake further research in this area. At the same time, it should be noted that by asking respondents about the most convenient forms of parcel collection offered and preferred by customers, the following results were obtained: the most preferred and currently chosen form of delivery is courier, the least personal collection (Table 1).

An interesting conclusion can be drawn with regard to the form related to the use of a mobile device. In terms of the assessment of 1–5—the first and second positions are dominant, they indicate that as many as 42.03% of the respondents

Table 1 The most convenient way to collect parcels purchased in the e-commerce system (in %)

	1	2	3	4	5
Courier	57.97	20.29	4.35	8.70	8.70
Parcel machine (self-collection point)	27.54	42.03	11.59	13.04	5.80
The reception point of the logistics company	2.90	13.04	34.78	33.33	15.94
Parcel on the move, gas station	1.45	11.59	33.33	30.43	23.19
Personal collection	10.14	13.04	15.94	14.49	46.38

1 Most comfortable most preferred, 5 least preferred, least comfortable form of parcel collection
Source: own elaboration

consider this form to be 2 for convenience and preferences, and as many as 27.54 for the first one. This may be due to the fact that the possibility of personal contact with the courier and the still existing belief that personal delivery by the courier is of a higher quality and more comfortable (no need to go anywhere for a parcel) still places the courier in first place. However, it is clear that this relatively new form of delivery (parcel machine) gains, both in terms of preferences, time and quality of delivery. The presentation of the possibilities of the new, open system, with the use of automatic devices from the collection of parcels, changes the opinion of potential customers about the ways of collecting parcels.

With the offered options for future delivery, they are usable for all suppliers and customers were proposed to deliver via:

- Courier (personal delivery to the place of residence)
- Self-collection point (parcel machine)
- Logistics company pick-up point (parcel box, post office)
- Package on the move, gas station
- Personal collection (from the seller)
- Open system with using of parcel lockers
- Drones
- "PaketButler"
- Multichannel

Preferences and possibilities for the development of such a concept were also asked. The results of the research indicate that the following possibilities are most often taken into account at the moment: courier, self-collection point (already in operation), parcel lockers (delivery box) and drones. Occasionally, it was pointed out that the collection points of a logistics company or a packet butler and multi-channel solutions were used.

Indication of the possibilities offered by the commonly available open parcel lockers system presented possible future solutions in the delivery of parcels within the "last mile". Interpreting the pilot studies, it is possible to say that in the future the delivery by courier and open parcel lockers system will have almost the same importance and so it can be concluded that the role of now operating parcel machines (one company as an owner) will decrease slightly, although still a large number of respondents, because as many as 56% indicated that this form will play a role in the future (putting its services on the first, second or third place). As far as further development of traditional courier services is concerned, the research shows that 47% of respondents still select on this form. At the same time, it puts it in the first place on an equal footing with courier services, at the same time indicating the highest percentage of respondents who indicated this form of service in the future on the third place (it may mean that other alternative choices will be more advantageous for customers in the future). It is also clear that the form of self-service devices is gaining importance, more than 56% of respondents indicated the currently functioning parcel machine as a form that they will choose in the future, including the fact that they did not indicate this solution in the first place, you are in the second position, you can guess, then, that the electors will either continue to use couriers, or will

Table 2 Which forms of parcel delivery will be the most preferred in the future

	in %		
N = 169	1	2	3
Courier (personal delivery to the place of residence)	33.33	10.14	30.43
Self-collection point (parcel machine)	26.09	34.78	17.39
Delivery box (open system)	33.33	43.48	18.84
Drones	7.25	7.25	14.49

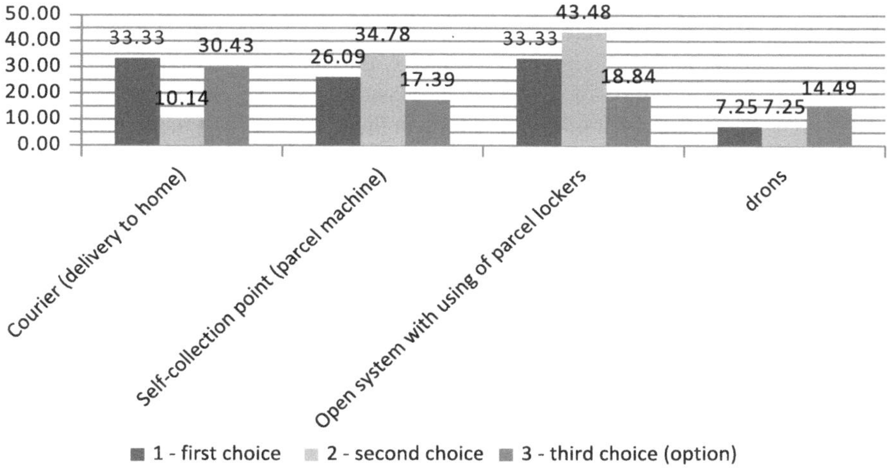

Fig. 4 Preferred forms of delivery in the future

convince themselves to a new form of business (Table 2). The new form of public courier box, whose added value, apart from servicing all bidders, is also speed, cost and delivery time, has been pointed out equally with couriers (it should be remembered that there is no system operating in this way yet, which may indicate a well-identified market need for the development of the courier services market and solving the last mile problem, and filling the market gap), in this case, as many as 91% of respondents opted for this form of delivery, placing this type of services on the first place (33.33%) and second place (42.48%). An interesting solution, already introduced on the American market, is the service of sent and last mile with the use of drones. It is interesting to note that the solution is also in the interest of respondents. Although a small percentage of respondents supported the solution, it is currently not available on the Polish market, which may mean that customers are interested in novelties and innovations.

The results of the pilot study indicate (Fig. 4) that there is growing interest in alternative solutions for handling and delivering parcels within the last mile. It is clear that customers are looking for new solutions that will give them more value. Such solutions should also be sought and offered by companies supplying goods.

This is related to the rapidly changing market and the growing importance of the e-commerce market. The graph only takes into account the answers given (no answer was taken, i.e. the choice of a given option).

4 Discussion and Summary

An important element in creating value for stakeholders is combining the concepts of different management strategies and new business models. In this case, we can talk about a combination of e.g. VBM (value based management) and supply chain management in the context of servicing the "last mile" and creating new solutions in this area. This means that one should look at the functioning of the company and the increase in its value in a multidimensional way, focusing on the concept of shareholders concerning meeting the needs of shareholders (company's shareholders) and the concept of stakeholders related to meeting the claims of all company's stakeholders, creating a platform for cooperation based on the balance of achieved goals in the perspective of long-term generation of positive cash flows. The concept of stakeholders assumes a much broader look at the objectives of the company, and the beneficiary of the created value is each stakeholder of the organization [36]. Therefore, these aspects can be considered in the context of, on the one hand, benefits for system builders and service providers, as well as customers. On the one hand, value is to be created for couriers, on the other hand, for customers using the system, and from yet another perspective, the last mile service is to be less harmful to urban space, which can also be considered as an aspect of added value. Maintaining positive relations with all stakeholders, including customers and offering them increased value, providing better service and solving problems arising in this area, as well as sustainable development will be a positive effect for all parties, simultaneously conditioning the development of the organization, creating new business models. At the same time, it should be pointed out that the desire to combine enterprise value management from the perspective of owners' interests with customer value management results directly from the strategic concept of the value chain. The key issue of designing business models is the mutual compilation of the organization's strategy, value chain and the resulting final proposal of value for the target customer. And for this purpose, a new model of "last mile" service can be used by the presented courier services. The basic element of an effective business model is the creation of unique value for customers and all stakeholders. While creating new business models, it is necessary to take into account creating value not only for enterprises (in the traditional approach) but also creating value for the customer and using his or her value. Therefore, it can be pointed out that customer value is the basis of company value and if technology enables the management of customer knowledge, then today's fundamental problems include: the approach to measuring customer value and the scope and source of acquiring, gathering and using customer knowledge to maximize their value for the company—building a strategy for increasing customer value [37]. In such a presented approach, it seems advisable and justified to extend this approach to all stakeholder groups, whose value understood as a set of behaviours and their mutual strategic interactions with the company may significantly become a source of growth in the value of a socially responsible company. An important role is played by the qualitative nature of stakeholder values, which locates the company in the appropriate market segment. Therefore, building a company strategy based on stakeholder value growth may now become an important element of effective and efficient management [38]. The increase in value for all stakeholders in the scope of the main assumptions resulting from the idea of delivery box service may result from the introduced innovations: process, product, technological and organizational, based on the use of the interface of economic exchange between business stakeholders, state and local government administration bodies and non-profit organizations and final customers. The result of such solutions, and generally the assumptions of the presented new business model, is the development of a new product or group of products and finding a potentially new market for them, at the same time creating a new element of the supply chain strategy and its links, including the final customers. Thanks to this approach, the value of stakeholders can significantly contribute to better identification and growth of the factors shaping the company's value, such as value drivers. At the same time, the most important barriers to the proposed model include: lack of interest of customers and companies that already owning infrastructure will not want to change distribution channels, lack of proven financial benefits, unreliability of the system based on information technology.

Pilot studies carried out indicate that the subject is interesting to customers and may constitute a new, interesting field of scientific consideration. The reflection is the first stage of research on the development of a new business strategy with a focus on creating added value for all stakeholders and the efficiency of the last mile while reducing the negative effects of logistics operations. The presented research results are an interesting study and consider for further consideration and conceptual development.

References

1. Moore, A.M.: Innovative scenarios for modeling intra-city freight delivery. Transp. Res. Interdiscip. Perspect. 100024 (2019)
2. Brown, J.R., Guiffrida, A.L.: Carbon emissions comparison of last mile delivery versus customer pickup. Int. J. Logist. Res. Appl. **17**(6), 503–521 (2014)
3. Revision of World Urbanization Prospects, UN, New York, (2018)
4. Raport E-Commerce w Polsce (2018), Gemius dla e-Commerce Polska. Retrieved, 31, 2019
5. de la Tore, J., Moxon, R.W.: E-commerce and global business: the impact of the information and communication technology revolution on the conduct of international business. J. Int. Bus. Stud. **32**(4), 617–635 (2001)
6. Lewicki, M.: E-handel w Polsce–stan i perspektywy rozwoju. Handel Wewnętrzny. **4**(375 tom I), 176–189 (2018)
7. Chaberek-Karwacka, G.: Teoretyczne kryteria kształtowania logistyki ostatniej mili i realne możliwości ich wykorzystania na obszarze największych aglomeracji w Polsce. Prace Naukowe Uniwersytetu Ekonomicznego we Wrocławiu. **383**, 13–25 (2015)

8. Weltevreden, J.W.: B2c e-commerce logistics: the rise of collection-and-delivery points in The Netherlands. Int. J. Retail Distribut. Manag. **36**(8), 638–660 (2008)
9. Ashraf, A.R., Thongpapanl, N.T., Spyropoulou, S.: The connection and disconnection between e-commerce businesses and their customers: exploring the role of engagement, perceived usefulness, and perceived ease-of-use. Electron. Commer. Res. Appl. **20**, 69–86 (2016)
10. Zhou, L., Wang, W., Xu, J.D., Liu, T., Gu, J.: Perceived information transparency in B2C e-commerce: an empirical investigation. Inform. Manag. **55**(7), 912–927 (2018)
11. Mahadevan, B.: Business models for Internet-based e-commerce: an anatomy. Calif. Manag. Rev. **42**(4), 55–69 (2000)
12. Chen, Z., Dubinsky, A.J.: A conceptual model of perceived customer value in e-commerce: a preliminary investigation. Psychol. Market. **20**(4), 323–347 (2003)
13. Savrul, M., Incekara, A., Sener, S.: The potential of e-commerce for SMEs in a globalizing business environment. Procedia Soc. Behav. Sci. **150**, 35–45 (2014)
14. Song, L., Wang, J., Liu, C., Bian, Q.: Quantifying benefits of alternative home delivery operations on transport in China. In: 2016 16th International Conference on Control, Automation and Systems (ICCAS), pp. 810–815. IEEE (2016)
15. Paredes, M.R., Barrutia, J.M., Echebarria, C.: Resources for value co-creation in e-commerce: a review. Electron. Commer. Res. **14**(2), 111–136 (2014)
16. Schumpeter, J.: Teoria rozwoju gospodarczego, p. 37. PWN, Warszawa (1960)
17. Anuszkiewicz, K., Marona, T.: Rola interesariuszy w rozwoju przedsiębiorstwa odpowiedzialnego społecznie, Rynek—Społeczeństwo—Kultura nr 1/2012, s. 31
18. Rutkowski, K.: Branżą przesyłek kurierskich, ekspresowych i paczkowych; wpływ na polską gospodarkę..Centrum Doradztwa i Ekspertyz Gospodarczych SGH, Warszawa (2011)
19. Kawa, A.: Analiza Rynku Kep W Polsce, GS1, Poznań (2017)
20. Zamkowska, S., Kogut, K.: Zakres i ocena oferowanych usług KEP przez wybranych operatorów logistycznych. Autobusy: technika, eksploatacja, systemy transportowe. **14**, 1099–1110 (2013)
21. Wang, X., Yuen, K.F., Wong, Y.D., Teo, C.C.: An innovation diffusion perspective of e-consumers' initial adoption of self-collection service via automated parcel station. Int. J. Logist. Manag. **29**(1), 237–260 (2018)
22. European Commission: Green Paper. An integrated parcel delivery market for the growth of e-commerce in the EU 698 final (2012)
23. Dieke, A.K., Bender, C., Campbell Jr., J.I., Cohen, R.H., Müller, C., Niederprüm, A., et al.: Main Developments in the Postal Sector (2010-2013). WIK Consulting, Bad Honnef (2013)
24. Morganti, E., Seidel, S., Blanquart, C., Dablanc, L., Lenz, B.: The impact of e-commerce on final deliveries: alternative parcel delivery services in France and Germany. Transp. Res. Procedia. **4**, 178–190 (2014)
25. Yuen, K.F., Wang, X., Ng, L.T.W., Wong, Y.D.: An investigation of customers' intention to use self-collection services for last-mile delivery. Transp. Policy. **66**, 1–8 (2018)
26. CBRE Research, https://www.cbre.com/research-and-reports [05.07.2020], Q4 (2016)
27. Vakulenko, Y., Hellström, D., Hjort, K.: What's in the parcel locker? Exploring customer value in e-commerce last mile delivery. J. Bus. Res. **88**, 421–427 (2018)
28. Grönroos, C.: A service perspective on business relationships: the value creation, interaction and marketing interface. Ind. Market. Manag. **40**(2), 240–247 (2011)
29. Djelassi, S., Diallo, M.F., Zielke, S.: How self-service technology experience evaluation affects waiting time and customer satisfaction? A moderated mediation model. Decision Support Syst. **111**, 38–47 (2018)
30. Ranieri, L., Digiesi, S., Silvestri, B., Roccotelli, M.: A review of last mile logistics innovations in an externalities cost reduction vision. Sustainability. **10**(3), 782 (2018); Chen, Q., Conway, A., Cheng, J.: Parking for residential delivery in New York City: regulations and behavior. Transp. Policy. **54**, 53–60 (2017)
31. Quak, H., Balm, S., Posthumus, B.: Evaluation of city logistics solutions with business model analysis. Procedia Soc. Behav. Sci. **125**, 111–124 (2014)

32. Collier, J.E., Sherrell, D.L.: Examining the influence of control and convenience in a self-service setting. J. Acad. Market. Sci. **38**(4), 490–509 (2010)
33. Lee, H.J., Lyu, J.: Personal values as determinants of intentions to use self-service technology in retailing. Comput. Hum. Behav. **60**, 322–332 (2016)
34. Lin, J.-S.C., Chang, H.-C.: The role of technology readiness in self-service technology acceptance. Manag. Serv. Qual. Int. J. **21**(4), 424–444 (2011) Theory of planned behaviour (Blut et al., 2016)
35. Oh, H., Jeong, M., Baloglu, S.: Tourists' adoption of self-service technologies at resort hotels. J. Bus. Res. **66**(6), 692–699 (2013)
36. Pachciarek, H., Szarek, M.: Kreowanie wartości dodanej dla interesariuszy-studium przypadku firmy handlowo-usługowej. Zarządzanie i Finanse. **15**(2, cz. 2), 293–307 (2017)
37. Dobiegała – Korona, B.: Klient kapitałem przedsiębiorstwa, Praca zbiorowa pod redakcją naukową B.Dobiegały – Korony i A. Hermana, Współczesne źródła wartości przedsiębiorstwa, Difin, Warszawa, s.94–95 (2006)
38. Jabłoński, M.M.: Kształtowanie modeli biznesu w procesie kreacji wartości przedsiębiorstw(Vol. 10). Difin. Oraz Jabłoński A, Jabłoński M. Watość interesariuszy jako kluczowy czynnik wartości organizacji odpowiedzialnych społecznie, Problemy jakości, 2008 (2013)

Convenience as a Factor Differentiating Air Transport Service

Dariusz Tłoczyński

Abstract Air transport is the branch with the fastest adaptation to the changing conditions of the internal and external environment. Carriers are constantly monitoring the market, examining passengers' needs and delivering modernized transport services. One of the factors shaping the service in passenger air transport is convenience. Although it is descriptive, it can be analyzed which factors and how they shape it, while also affecting the transport service. The purpose of the work described in this way was subjected to the critical analysis of available literature and the analysis of marketing research carried out at Polish airports on the factors affecting the development of air transport.

On the basis of the presented research results, it was found that there are three levels describing the present state of diversification of the air transport service. Undoubtedly, such analysis of research on the issue of convenience should be carried out not only by academic centers, but also by institutions involved in transport activities. The conclusions from the conducted research can be the basis for a broader analysis of the selected research problem.

Keywords Convenience · Air transport · Legacy carrier · Low-cost carrier · Transport service · Passenger · Airports

1 Introduction

In 2017, over four billion people from around the world used air transport. Such a large number of passengers, as well as the features of air transport, require further actions that can be taken to meet the needs of potential travelers on the one hand and respond to market changes and passenger expectations on the other hand. Experts responsible for air traffic forecasts working for major aviation equipment manufacturers estimate that the passenger market will grow at a rate of 6% per year.

D. Tłoczyński (✉)
Faculty of Economics, University of Gdańsk, Gdańsk, Poland
e-mail: d.tloczynski@ug.edu.pl

M. Suchanek (ed.), *Transport Development Challenges in the 21st Century*, Springer Proceedings in Business and Economics,
https://doi.org/10.1007/978-3-030-50010-8_6

Prospects related to market development lead to the development of transport services. The factor that accelerates the development is undoubtedly scientific and technological progress and globalization and integration processes. In addition, the factors that stimulate the development of the aviation market are: development of the wealth of the society, development of regions, inter-branch competition, and above all a change in the expectations of buyers—consumers of services in passenger air transport.

By modernizing the transport service, carriers conduct a series of research on passengers' expectations, examine transport needs and behaviors. The results of these analyzes, however, are not found in official reports of transport companies due to the competitive threat. Therefore, there is a need for such analyses to be carried out by scientific communities dealing with air transport issues. Therefore, research was undertaken on one aspect of the transport service—convenience. For this purpose, factors influencing the convenience of transport services were determined and analyzed using proprietary marketing research carried out at Polish airports. The conclusions from the presented research should form the basis for further work related to the issue of the air transport services market.

2 Nature of Convenience

The air transport services market is a dynamic element of the global economy. In the world in 2017 carriers transported over four billion passengers, which means that there was an increase in the number of passengers carried by 7.1%. The transport performance increased by 7.9% and passengers were offered almost 42 million flights [1–3]. In the last few years on the market, we have been observing a continuous increase in the number of people purchasing air transport services. This, in turn, leads to increased demand and offer of transport services.

Air transport services are one of the most important activities of the European and global economy, which are important for both economic development and the progress of integration processes [4]. Safety, on-time, reliability, convenience in terms of airport proximity or seat availability, frequency of departures, in-flight cabin services, ground services including ticketing and baggage handling, aircraft type, and even the carrier's image are part of the airline product.

Authors studying air transport issues [5–7] indicate that the market attractiveness of the air service, along with its high quality and satisfying price, are:

- A network of convenient connections
- The offered flight schedule and coordinated in time arrivals and departures in various relations
- Type of aircraft used
- Accessibility to differentiated (in terms of the level of quality) travel conditions
- Comprehensive and efficient service for service users at airports

In addition to the features listed above, J.G. Wensveen [8] also indicates:

- Catering on board of aircraft
- Providing adequate entertainment on board aircraft (long-distance routes) and at airports (play areas for children)
- Sales in duty free shops, on board aircraft
- The possibility of transporting luggage
- Bilateral agreements, alliances
- Other amenities

The features of the air service should be synchronized with the goals and production capabilities of carriers, airports, market conditions of their macro- and micro-environment.

The distinctions of the air service are achieved by air carriers by:

- Adaptation of the air service to selected market segments (legacy, hybrid and low-cost carriers)
- Effective use of owned air fleet and management of the network of connections [9]
- Launching new functions of the service on the market—additional functions
- Increasing the level of quality of offered services
- Use of loyalty programs and branding
- Partnership and cooperation between aviation entities

One of the many factors that affects the choice of aircraft as a means of transport is convenience. In the most general way, convenience is defined as a type of good that is purchased often by a consumer without involving much exertion or consideration [10]. However, such an interpretation of convenience cannot be directly referred to a transport service. The Oxford English Dictionary [11] defines convenience as "...the state of being to proceed with something without difficulty ... the quality of being useful, easy or suitable for someone... a thing that contributes to an easy and effortless way of life...."

Claffey [12] was the first to try to define convenience in relation to the transport service. Despite the changes in the progress in science and civilization related to the development of transport, the concept of convenience (adapting personal plans and habits of passengers to use transport services and overcoming difficulties related to the provision of services) is still valid. On the other hand [13], they define convenience as a certain individualism related to expectations in relation to the provision of a transport service. According to this team, convenience refers to the efficiency and effectiveness with which a person can be transported from point A to the destination. The analysis was also made in relation to access to the transport vehicle, waiting for the vehicle, entry and exit, and time spent in the vehicle.

In turn, OECD [14] specifies more precisely the concept of transport users' convenience in terms of comfort, reliability, and safety. In addition, convenience should be analyzed by assessing the effectiveness of the use of a transport vehicle. Berry et al. [15] identified five types of convenience:

- Decision about choosing a transport means
- Availability

- Freedom of transactions
- Expectations regarding the performance of the transport service
- Benefits from the implementation of a comfortable trip

at the same time stating that convenience is characterized by discretion. And there is no doubt about this statement.

In Polish literature on transport economics, convenience is descriptive and it should be agreed with Barry and Anderson [16] that it is characterized by discretion. Convenience as the expectation of passengers directed to the organizers and carriers is considered as:

- Reliable
- Punctual
- Safe service
- Simplicity of the tariff system
- Freedom to purchase a ticket
- Understandable system of communication information
- Freedom to reach the stop (number of pedestrian crossings, tunnels, architectural obstacles)
- Legibility of timetables
- Legibility and aesthetics of marking (lighting) of stops and vehicles
- Ease of getting on and off
- Driving comfort (air conditioning, noise level, lighting, seating position)
- Service culture
- Luggage space
- Access to catering services [17]

3 Convenience Management as an Element of the Transport Service

Currently on the market of air transport services, there are two main models of passenger transport by regular air transport. However, one should point to the model combining selected features of low-cost and legacy carriers—a hybrid model. Such a large differentiation of the features of the air service results from the adjustment of the offer to the selected market segment. Passengers treat the selected offer of transport services as something special—a formulated offer for the buyer that other airlines could not offer.

Differentiation of the transport service takes place by, among other, convenience. Although it is discretionary and descriptive, carriers use different ways of influencing the purchase of air services. The comparison of factors affecting the convenience of transport services offered by legacy, hybrid, and low-cost carriers is presented in Table 1.

Table 1 Comparison of factors affecting convenience in various business models of air carriers

Feature of air service	Legacy carriers	Hybrid carriers	Low-cost carriers
The method of booking a ticket	Carriers' websites, mobile applications, websites of air brokers, offices of carriers' representations	Carriers' websites, mobile applications, websites of air brokers	Carriers' websites, mobile applications, websites of air brokers
The method of check-in	Carriers' websites, mobile applications, web-point	Carriers' websites, mobile applications	Carriers' websites, mobile applications
On-board and catering service	Free	Payable	Payable
The quality of services provided	Good	Good	Poor
Space in the cabin, per 1 passenger	Huge	Little	Little
A seat in the passenger cabin	Marked on the ticket, the ability to choose a specific seat free of charge	Marked on the ticket, the ability to buy the selected seat	Marked on the ticket, the ability to buy the selected seat
Class of the carriage	Economy, business, first	Economy, business	No
Change of date	In cheaper tariff systems there is no such possibility, in more expensive there is such a possibility	For extra charge	For extra charge
Aircraft	Several types, a few, a dozen or so years old	One type, several years old	One type, several years old
Accessibility of the airport	Central, main, and regional airports, "point to point" connections, transit connections	Main, regional, and local airports, tourist cities, only "point to point" connections	Regional and local airports, tourist cities, most airports distant from city centers, only "point to point" connections
The length of the route	It doesn't matter	Continental and selected intercontinental routes	Continental connections and selected extra-continental destinations

(continued)

Table 1 (continued)

Feature of air service	Legacy carriers	Hybrid carriers	Low-cost carriers
Amenities	Cooperation with travel agencies, hotels, car rentals, the possibility of carrying unusual cargo, luggage, diversified baggage policy, luggage most often the free most-used dimension of free luggage 50 × 40 × 20	Cooperation with travel agencies, hotels, car rentals, the possibility of carrying unusual cargo, luggage, a varied baggage policy. Luggage depending on the carrier payable or free, the most commonly used dimension of free luggage 50 × 40 × 20	Cooperation with travel agencies, hotels, car rentals, the possibility of carrying unusual cargo, luggage, diversified baggage policy, paid luggage, the most frequently used dimension of free luggage 30 × 20 × 40
Catering	Free, the possibility of buying products from the on-board catalog	Payable, the possibility of buying products from the on-board catalog	Payable, the possibility of buying products from the on-board catalog
Loyalty programs	Yes	Yes	No
Belonging to the organizations and alliances	Yes	No alliances, there is the European Low Fares Airline Association (ELFAA)	No alliances, there is the European Low Fares Airline Association (ELFAA)
Access to the air service	High frequency, timing, and ticket coordination with the use of transit	High frequency dependent on final destination and demand	Frequency dependent on final destination and demand
Check-in time at the airport	Long	Very short	Very short
Transport on the route airport—aircraft	Most often the jet bridge, airport bus	Most often on foot, possibly the jet bridge or airport bus	Most often on foot, possibly the jet bridge or airport bus

The division into legacy and low-cost carriers as well as growing competition in terms of inter-branch and aviation market necessitates constant adjustment of transport services to the changing passenger preferences. Hybrid carriers, while carrying out transport activities, enrich the basic service with free or partially payable additional services offered, however, operating a low-cost model of functioning. Hybrids are also created by legacy carriers, which reduce operating costs and implement methods tested by low-cost carriers [18].

Carriers striving to offer a whole package of services introduce a number of modifications, most often IT and technology. An example is the factor describing the convenience is mobile first (preferring the information command on smartphones, not computers). It is already happening and will continue. Passengers connected to the Internet expect information available on demand, access to services at your fingertips. Despite the fact that the purchase, check-in, and check-out are now standard, passengers expect information about the number, luggage number, information about the accessibility to the city center (airport link), possible delays, commercial and gastronomic offer, or even the best route to reach the gate.

The presented features of the transport service are instruments that distinguish air carriers' services from the competition. Although the parameters describing convenience are very similar to the presented market segments, from the point of view of passengers they are perceived as elements diversifying the air service.

4 The Assessment of Convenience Based on Conducted Scientific Research

In 2013, a study was carried out to determine the impact of factors on the decision to choose an air carrier. The study was conducted on 1% of air traffic, taking into account the type of business (LCC or Legacy).The research was carried out at almost all airports with the exception of Modlin, and Olsztyn—during the survey the airports were not operating. The method used in the study was personal interview with passengers taking a regular travel. Respondents were selected on the basis of stratification based on the number of passengers, taking into account the specificity of air traffic in each airport.

Based on the research conducted at Polish airports, passengers assessed the factors describing convenience for 57% of passengers, convenience is a factor encouraging to travel by air and for 38% it is a factor that is not important in the decision-making process on air travel (Fig. 1).

In the case of passengers choosing an air carrier, a similar meaning of convenience can be observed with regard to the selection of the air service. For 51% of respondents, convenience is a factor encouraging the choice of carrier (this is the

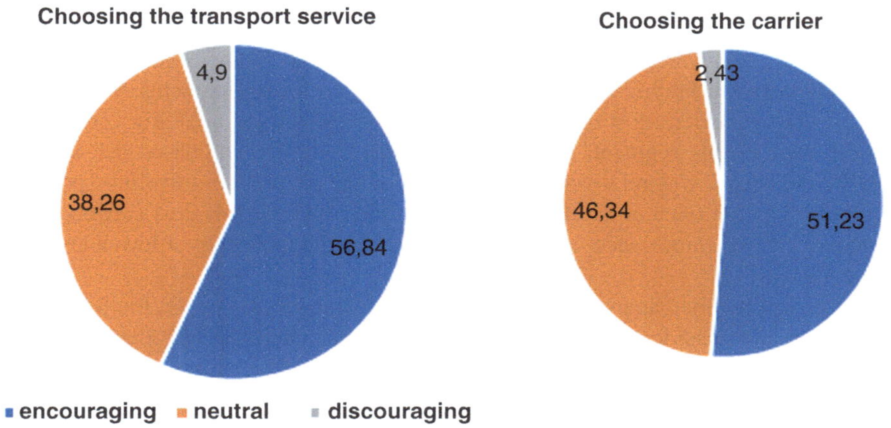

Fig. 1 The role of convenience in the process of choosing an air transport service and an air carrier. (Source: own marketing research)

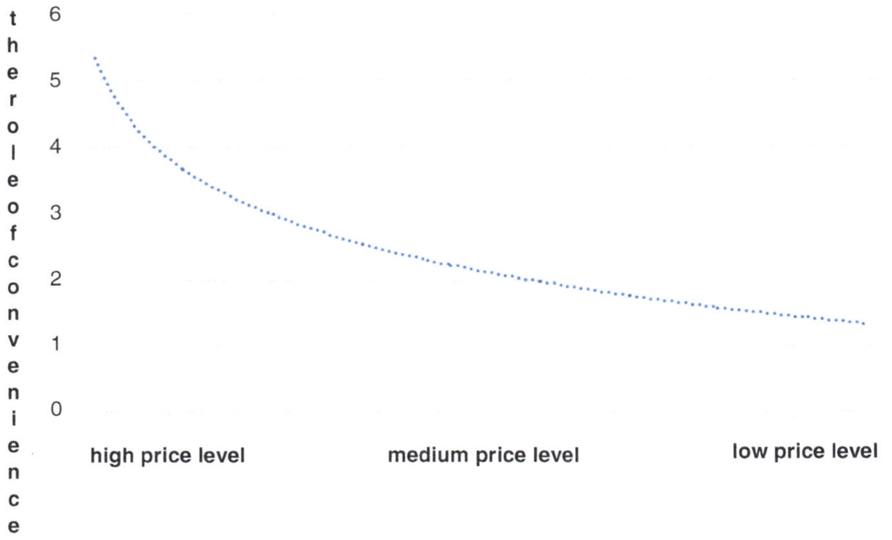

Fig. 2 The role/importance of convenience depending on the ticket price. (Source: own marketing research)

majority of travelers who travel legacy air carriers), while for 46% comfort does not affect the choice of carrier. Such a division of respondents' answers indicates that convenience is not a significant determinant of the choice of transport service and air carrier.

For most travelers the convenience of air travel is something natural, adequate to the way of travel. However, the convenient meaning depends on the ticket price (Fig. 2).

Analyzing the research results, there was a relation between the degree of satisfaction with convenience and the price of the air ticket. Higher expectations and consequently the feeling of convenience were expressed by passengers traveling with a legacy carrier, most often people having business trips, traveling very often. In the case of this segment, the price for a ticket, usually purchased a few days before departure, was relatively high. In the case of passengers traveling sporadically, most often once or several times a year and choosing low-cost carriers (low price level), the importance of convenience was characterized by a lower level of acceptability.

Making a detailed analysis of convenience as an air service factor, the following was taken into consideration (Fig. 3): directness, ticket purchase and check-in using electronic mobile devices, convenient departure and arrival times, catering, on-board service, terminal functionality, satisfaction with the time of arrival at the airport, offer and price level of services at airports, purchase of additional services.

The greatest impact on the perception of the convenience of the transport service in the opinion of the surveyed passengers was directness, satisfaction with the time

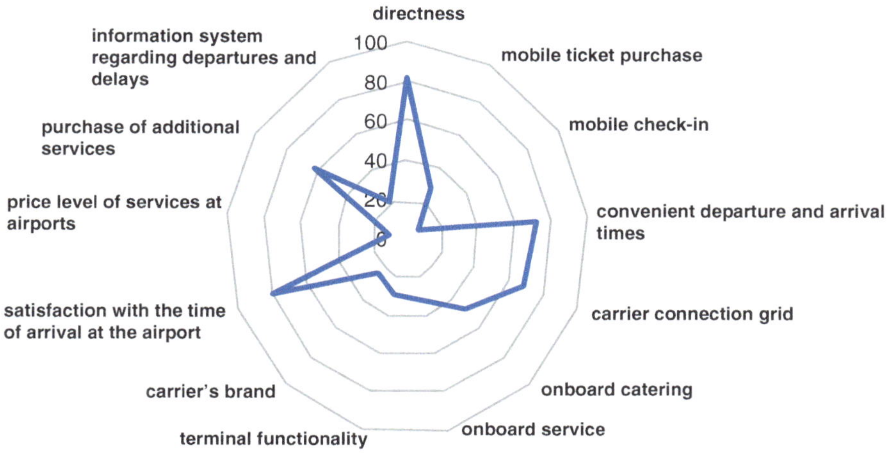

Fig. 3 The impact of selected factors on the perception of transport service convenience. (Source: own marketing research)

of arrival at the airport, the network of air connections, and convenient arrival and departure times. However, passengers indicated those areas that require analyses to play an important role in shaping the convenience of transport services in the near future. Respondents included there:

- Price level of services at airports—due to the specificity of the industry, places of sale, requirements. One should not expect in the near future a reduction of prices of products and services at airports.
- Mobile check-in—this factor is becoming more and more attractive over time. Currently, it is becoming more and more common to make check-in using mobile applications, however, online check-in is an obstacle.
- Mobile ticket purchase—in the era of progressing digitalization, a smartphone or tablet becomes an increasingly popular means of purchasing travel documents. As in the case of check-in, the competition is the purchase of tickets via the Internet.
- Information system about flights, gates, delays of aircraft—at the moment it is not fully available in mobile applications, passengers must use information boards located in terminals.
- Carrier's brand and terminal functionality—passengers do not choose an airport deciding to start a trip, the purchase of a flight service is accompanied by a whole package of services related to transport, i.e., price, departure times, availability of the region, carrier's connection grid. The brand can be a factor enriching the air service only in the case of competition of two similar carriers offering similar conditions of carriage—but this is just a theory.

5 Summary

The transport service due to its characteristics is assessed by carriers representing various business models. Over the last dozen or so years, we have been observing the process of adapting the transport service to the changing and diverse needs of passengers. Convenience, although it is discretionary and descriptive, is a feature shaping transport services.

With regard to air transport, in the light of the conducted analyses, convenience consists of the directness, the method of ticket purchase and check-in using electronic mobile devices, convenient departure and arrival times, catering, on-board service, terminal functionality, satisfaction with the time of arrival at the airport, offer and price level of services at airports, purchase of additional services, but also space in the passenger cabin—distance between seats, service level in various market segments, loyalty program, the possibility of using executive lounge saloons—only selected passengers. In summary, there are many factors describing convenience.

On the basis of research carried out at Polish airports, it was found that the greatest impact on shaping the convenience of the transport service is directness, convenient departure and arrival times as well as time to reach the airport, and thus parameters affecting the total travel time. Air transport services, regardless of the segment served, are characterized by similar parameters in terms of region availability, rolling stock used, aesthetics of means of transport, and level of service quality.

On the other hand, the area that will be revolutionized and evaluated by passengers in the near future is the use of mobile devices in the entire transport service process.

References

1. World Airline Report. Air Transport World (2018)
2. Annual Review 2018. IATA (2019)
3. Economics Performance of the Airline Industry. IATA (2019)
4. Tłoczyński, D.: Konkurencja na polskim rynku usług transport lotniczego, pp. 124–127. Gdańsk University Press, Gdańsk (2016)
5. Czownicki, J., Kaliński, D., Marciszewska, E.: Transport lotniczy w gospodarce rynkowej, pp. 48–49. Warsaw School of Economics, Warsaw (1993)
6. Cheng-Lung, W.: Airline Operations and Delay Management, p. 7. Ashgate, Farnham (2010)
7. Ashford, N., Stanton, H.P.M., Moore, C.A.: Airport Operations. McGraw Hill, New York (2012)
8. Wensveen, J.G.: Air Transportation. A Management Perspective, pp. 263–265. Ashgate, Farnham (2007)
9. Belobaba, P.P.: The airline planning process. In: Belobaba, P.P., Odoni, A., Barnhart, C. (eds.) The Global Airline Industry, p. 154. Wiley, Chippenham (2010)
10. Business Dictionary. www.businessdictionary.com. Accessed 15 Apr 2019
11. The Oxford English Dictionary. Oxford University Press (2016)

12. Claffey, P.J.: User criteria for rapid transit planning. J. Urban Plan. Dev. Division ASCE (1964)
13. Stopher, et al.: Toward the development of measures of convenience for travel modes, Transportation Research Board, National Research Council, National Academy of Sciences USA. Sage Publishing (1974)
14. Wardman, M.: Valuing Convenience in Public Transport. Roundtable Summary and Conclusions. International Transport Forum. Discussion Paper No. 2014-02 2014. OECD (2014)
15. Berry, L.L., Seiders, K., Grewal, D.: Understanding service convenience. J. Market. **66**(3), 1–17 (2002)
16. Anderson, R., Condry, B., Findlay, N., Brage-Ardao, R., Li, H.: Measuring and Valuing Convenience and Service Quality: A Review of Global Practices and Challenges from Mass Transit Operators and Railway Industries. OECD/ITF Discussion Paper 2013-16. Prepared for OECD/ITF Roundtable on Valuing Convenience in Public Transport. Paris, 12–13 September (2013)
17. Koźlak, A.: Ekonomika transport. Teoria i praktyka gospodarcza, pp. 12–16. Gdańsk University Press, Gdańsk (2007)
18. Taneja, N.K.: Simpli-Flying, p. 44. Ashgate, Aldershot (2004)

Shaping Levels of City Bus Prices as a Result of Tenders

Aleksander Jagiełło

Abstract The article attempts to verify the research hypothesis formulated by the author stating that the size of the tender (understood as the number of once purchased vehicles) negatively affected the final unit price of the purchase of a city bus. Confirmation of the hypothesis would testify to the potentially greater cost-effectiveness of functioning of large carriers on the urban transport market. In order to verify the above hypothesis, the current situation of the city bus market in Poland, the criteria used in tender procedures and the results of primary research on the correlation between the size of the tender and the unit price that MAXI class buses in Poland had as a result of tender procedures were presented. In the final part of the article, the author presented the discussion and conclusions from the conducted research.

Keywords Rolling stock investments · Tenders for city buses · Prices of city buses · Tender criteria

1 Introduction

In microeconomics, economies of scale are the cost advantages that enterprises receive due to their scale of operation. The economies of scale occur when, together with the increase in the production volume of the enterprise, the average costs per unit of production decrease [1, 2, 15, 16]. This is because, as production volumes increase, long-term average total costs are reduced (fixed costs are distributed to a growing number of goods or services produced). As the analyses made so far on the intensity of this phenomenon on various markets show, the scale of the economies of scale on the level of average costs is particularly high in the automotive industry [3, 4]. In addition, as the volume of production increases, the positive effect of

A. Jagiełło (✉)
Department of Transportation Market, Faculty of Economics, University of Gdansk, Gdansk, Poland
e-mail: aleksander.jagiello@ug.edu.pl

M. Suchanek (ed.), *Transport Development Challenges in the 21st Century*, Springer Proceedings in Business and Economics,
https://doi.org/10.1007/978-3-030-50010-8_7

specialization and learning increases. This effect results from the greater productivity of the workforce with greater experience in the production of a given goods or service [5, 6, 17].

Considering the above conditions regarding the occurrence of economies of scale in the automotive market, the article attempts to verify the hypothesis that the increase in the size of the tender (understood as the number of once purchased vehicles) negatively affected the final unit price of the purchase of a city bus. The verification of the research hypothesis was used to verify the thesis saying that the size of the enterprise providing public urban transport services, and thus the size of tenders submitted and settled, influences the amount of the company's own costs (assuming the vehicle's own costs as a carrier). The impact of the level of bus purchase costs on the level of the own costs of a company providing public transport services is particularly evident during the calculation of the unit cost of public transport. The price of a new vehicle in addition to the annual depreciation charge is one of the component costs related to depreciation. Depreciation costs, on the other hand, constitute one of the direct costs of the operation of public urban transport [7].

2 The Market of City Buses in Poland in the Light of Tender Allotments

Currently, about 12,000 city buses and over 200 trolleybuses are in operation in Poland [8]. Despite constant rolling stock investments, which are carried out by carriers providing public transport services, there are no significant changes in the age structure of used buses throughout the country. In Poland in 2018, over 1000 new city buses were registered, which means a significant increase in the number of new city bus registrations compared to previous years (2016—723, 2017—771). According to the data of the Chamber of Commerce Municipal Transport presented in Fig. 1, in 2017 the registration of new city buses resulted in the recovery indicator of public transport buses which is a share of purchases in inventory amounting to 5.5 (IGKM in its methodology takes into account the purchase of not only new vehicles but also used vehicles) [9].

The purchase of new vehicles and hence the increase in the recovery indicator of rolling stock results directly in the condition and age of the Polish fleet of city buses. As shown by the data presented in Fig. 2, despite the yearly rolling stock investment, the percentage of buses used in Poland no older than 6 years oscillate between 25 and 30% with a downward trend (32%—2014, 31%—2015, 27%—2016, 25%—2017). Moreover, throughout the entire period under examination from 2000 to 2017, buses older than 10 years accounted for more than 40% of the entire fleet.

To improve the condition of the Polish fleet of city buses and hence the level of quality of the public transport services provided, it is necessary to purchase new vehicles. These purchases take place through tender procedures, most often in an

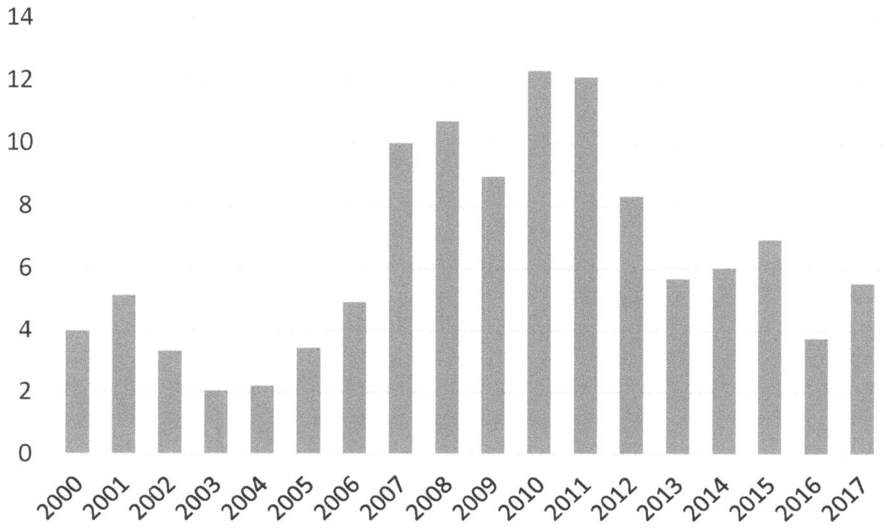

Fig. 1 Recovery indicator for urban transport fleet (calculated as the share of purchases in the inventory) in the years 2000–2017. (Source: IGKM, 2018)

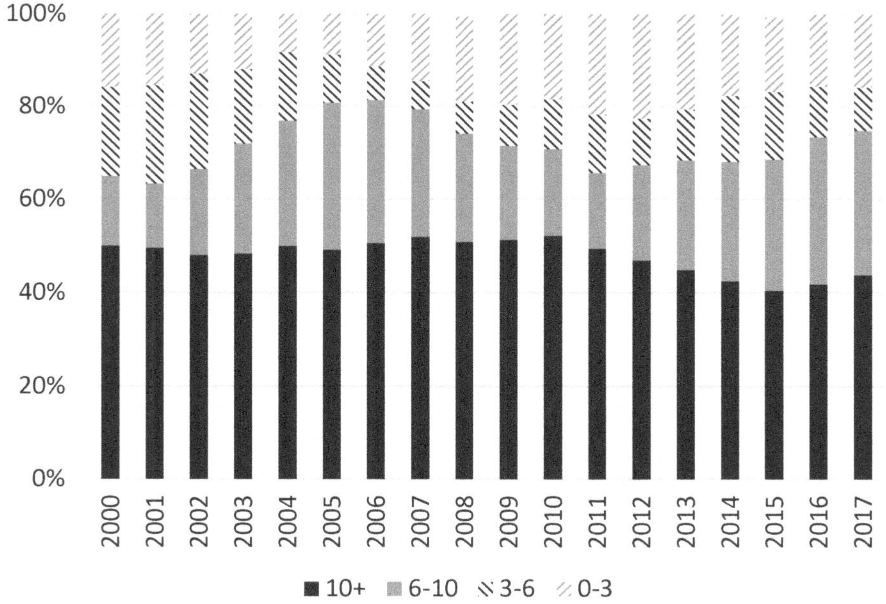

Fig. 2 Age structure of public transport buses in Poland (in %). (Source: IGKM, 2018)

unlimited procedure (in which all interested manufacturers of city buses can participate).

3 Criteria Used in Tender Procedures for City Buses

Through amendments to the law (i.e., the Act of 29th August 2014 amending the act—Public Procurement Law, Journal of Laws of 2014, item 1232) and the Act of 22nd June 2016 amending the Act—Public procurement law and some other acts (Journal of Laws of 2016, item 1020), the legislators limited the selection of the best offer by the contracting authority only on the basis of the lowest price criterion, as shown in Fig. 3, in the majority—68% of tendering procedures for city buses in Poland in 2018 were applied at least two non-price criteria.

Among the applied non-price criteria in tenders for city buses, the highest importance was given in 2018 to the technical parameters of the ordered vehicles (while analyzing individual tenders a large range of this criterion is visible) and conditions for vehicle warranty and service. The environmental aspects of rolling stock, technical readiness, fuel and electricity consumption, comfort and safety offered by the vehicle, prices of spare parts, and unification of engines of delivered buses were also assessed (Table 1).

Figure 4 presents data on 91 tenders for city buses with allotment in 2018. According to them, the price of city buses corresponded on average to 62% of the total offer evaluation submitted by the rolling stock producers for the tender (most often, in 77% of cases the price was 60% of the total offer evaluation price).

In conclusion, it should be noted that the price offered by vehicle manufacturers is still (though to a lesser extent than before the 2014 and 2016 changes) a decisive criterion for the selection of offers in tender proceedings for city buses in Polish conditions.

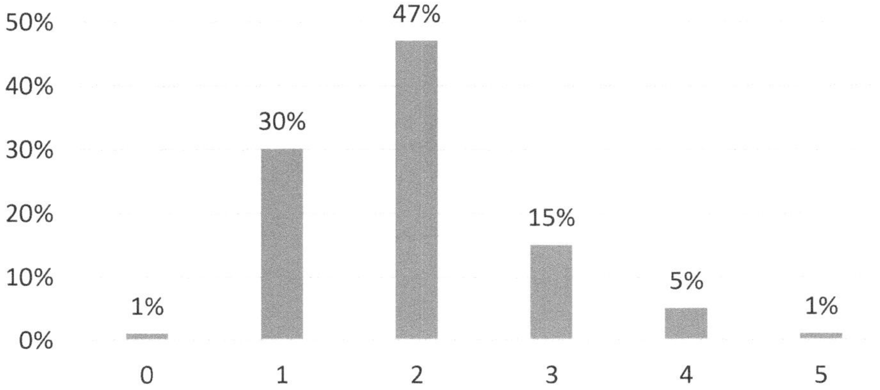

Fig. 3 The number of non-price criteria applied in tenders for city buses in 2018. (Source: own study)

Table 1 Importance of non-price criteria in tender procedures for city buses in 2018 (in %)

Criterion	Measure				
	Minimum	Mean	Maximum	Median	Dominant
Technical parameters	6	30	42	30	40
Terms of warrantee and service	4	11	30	10	10
Fuel/energy consumption	3	8	30	5	5
Delivery date	2	8	40	5	5
Ecology	5	12	35	9	10

Source: own study

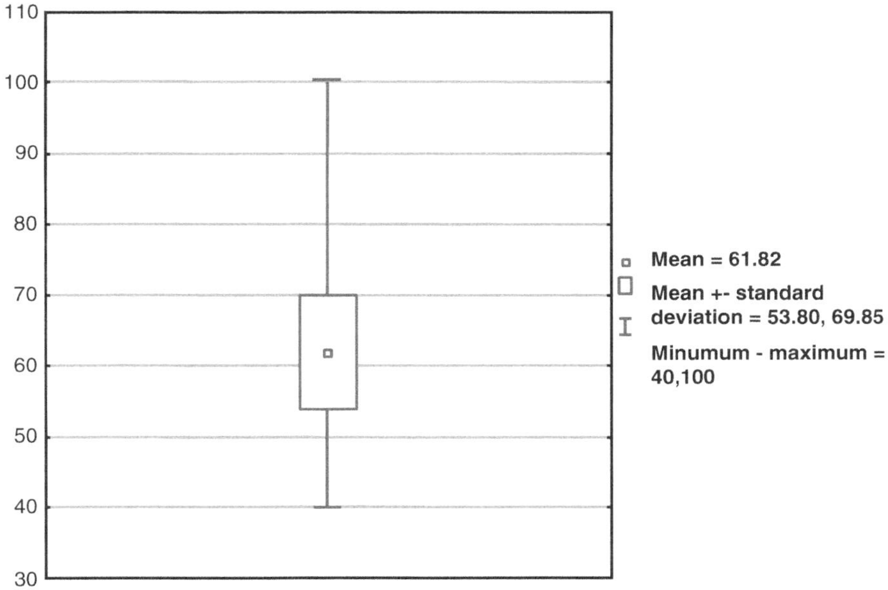

Fig. 4 Percentage of points given for the price in tenders for city buses in 2018. (Source: own study)

4 Results of the Analysis of the Impact of the Size of the Tender on the Unit Price of a City Bus

The data on the results of 180 tenders for 12 m single-stage city buses were subject to the study. These tenders were organized by carriers providing public transport services in Polish cities. The time period of the study covered tenders that took place in 2011–2018, and as a result 1736 city buses were purchased. Only tenders concerning 12 m buses of the MAXI class were subject to the study, because these buses constitute the most popular and the largest segment of city buses in Poland and their price significantly differs from the price of vehicles of other classes (MINI—below 8 m, MIDI—9–10 m, MEGA—13–15 m, and articulated buses). In addition, MAXI class buses are on offer from the largest number of rolling stock manufacturers, which has a direct impact on the number of tenders, and thus on the intensity of competition and the price level of the vehicles themselves. In order to obtain reliable results, only conventional buses driven by a diesel engine were tested, because their market price clearly differs from the prices of buses powered by LPG/CNG, electricity, or hybrid buses. Thus, shaping the price level of city buses powered unconventionally as a result of tenders requires separate analyses. For the reliability of the results, all bus prices for which tenders were settled in 2011–2017 were adjusted for inflation.

Based on the conducted analyses, the results of which are presented in Fig. 5, it should be stated that there is no negative correlation between the size of the tender and the unit price of a conventional city bus obtained as a result of the tender (coefficient of correlation = 0.21). In addition, the volatility of the price of city buses is only 4% explained by the volatility of the size of the tender (coefficient of determination = 0.044). The results of the regression analysis showed that size of the tender for 12 m buses in Poland has a statistically significant effect on the price of city buses ($p = 0.00481$). The estimated regression parameter b was equal to 2166.

In order to minimize the impact of other variables on the reliability of the obtained results, the impact of the tender size on the price of city buses in the case of tenders won by one producer (in the scope of one model) was also analyzed.

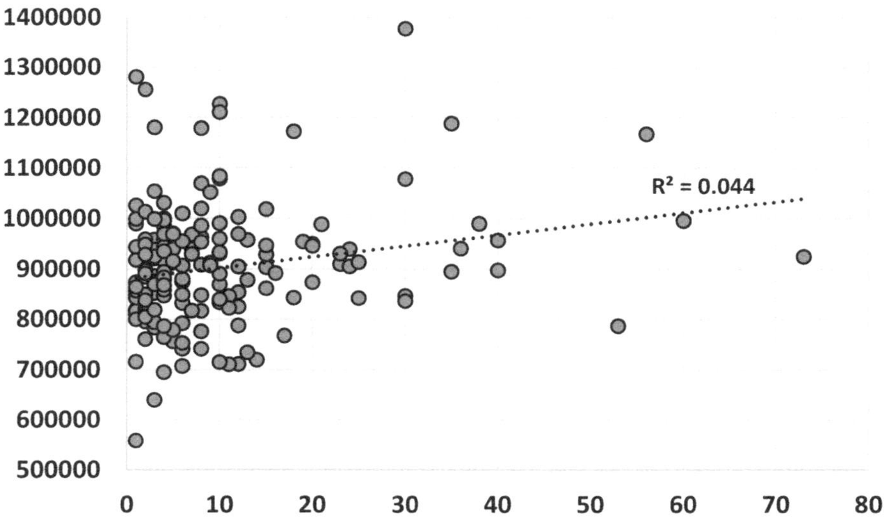

Fig. 5 Scatter chart—unit price of the bus in relation to the number of buses ordered in tenders in 2011–2018. (Source: own study)

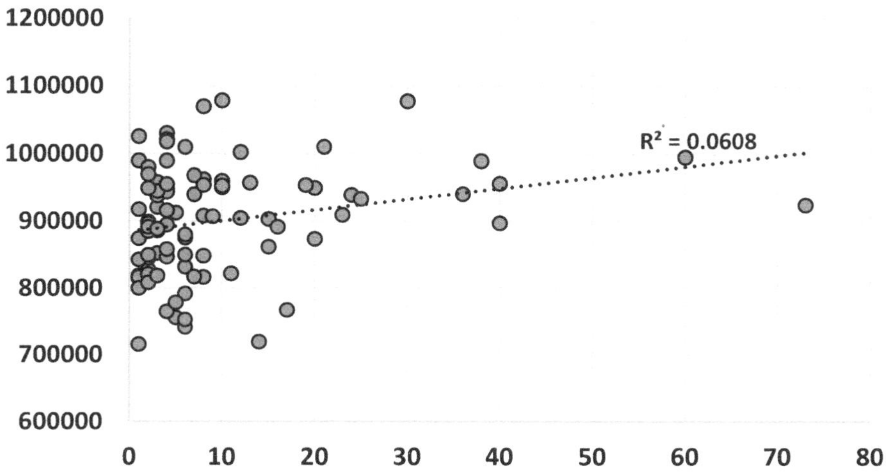

Fig. 6 Scatter chart—unit price of the Solaris Urbino 12 bus in relation to the number of buses ordered in tenders in 2011–2018. (Source: own study)

Thus, 95 tenders (as a result of which 913 buses were purchased) were subjected to analysis for 12 m city buses in which Solaris Bus & Coach SA won by offering Urbino 12 buses (the most popular city bus in Poland). The results of the analysis are shown in Fig. 6. In this case, there was also no negative correlation between the size of the tender and the unit price of Urbino 12 city bus obtained as a result of tenders (coefficient of correlation = 0.16). According to the analyses carried out, the volatility of the price of Solaris Urbino 12 buses is explained only in the 6% variation in the size of the tender (coefficient of determination = 0.06), which is almost the same as for the entire 12-meter bus segment. The results of the regression analysis showed that size of the tender for Urbina 12 buses in Poland has a statistically significant effect on the price of Urbina 12 buses ($p = 0.0185$). The estimated regression parameter b was equal to 1615.

The results of regression analysis and the value of correlation and determination coefficients obtained during the analysis mean that the confirmation of the hypothesis stating that the tender size (understood as the number of one-time purchased vehicles) had a negative impact on the final unit price of the city bus purchase was not obtained. Thus, it was unsuccessful to verify the thesis that the increase in the number of tenders submitted and allotted reduced the company's own costs (assuming the vehicle's own costs as a carrier) related to the purchase and operation of city buses.

5 Discussion

The study and analyses indicate that in the case of the automotive market, economies of scale are a particularly important factor influencing the final price of offered vehicles, and hence are an important factor affecting competition between manufacturers that form the supply side of the automotive market [3, 10, 11]. It results from high costs incurred for research and development of new models of vehicles put into operation, costs related to the development of modern technologies (among others aimed at reducing harmful substances emissions and reduction of fuel consumption), and high costs of creating production lines and their modernization in terms of production of new models and application of new technologies. Despite this, as a result of analyses carried out in this article, it was not proven that the producers of city buses are willing to reduce the unit price level in the case of a single order of a larger number of vehicles (through the tender procedure). Since in Poland so far no analyses have been carried out in the scope presented by the author in the article, it is impossible to confront the obtained results with historical results (from previous years). It is also impossible to say whether the received results were not affected by the exceptional situation on the city bus market. The uniqueness of the state of the Polish city bus market in the analysed period was related to the high level of demand caused by, among other things: European Union funds spent on the development of public transport, national and European policy aiming to protect the natural environment, progressing trends regarding the change of power supply for city buses and operation undertaken in the scope of changing the current modal split for the benefit of greater use of public transport. All the above factors had a positive effect on the volume of demand and thus discouraged the rolling stock producers to compete on price.

6 Conclusions

On the basis of the analyses presented in the article, it can be stated that in Polish conditions, the size of the announced tender is not a factor allowing to reduce the unit costs of buying a city bus. Thus, the research hypothesis was not confirmed at the time of the analysis, saying that the size of the tender (understood as the number of one-off vehicles purchased) affected the final unit price of the purchase of a city bus. Consequently, the thesis that the larger enterprise providing public urban transport services (and therefore the bigger tenders notified and resolved) the lower the costs of purchasing city buses, and hence the own costs of the enterprise (assuming the own vehicle costs as a carrier) was not confirmed. It should be noted, however, that the cost of buying a city bus is only about 20–30% of the total cost of its life cycle (this value is different in various studies due to the analysis method used, variables taken into account, and the quality of input data) [12]. Other important elements of the bus life cycle costs are the costs of vehicle maintenance (which include labor costs and parts and materials) and energy costs (liquid fuels, gas, or electricity). The amount of individual costs included in the life cycle cost of a bus is strongly differentiated between diesel-fueled, hybrid, CNG/LNG, and electric [13, 14, 19, 20]. The mileage that the vehicle will overcome during its operation also influences the differentiation of the individual components of the bus life cycle costs [12, 18].

References

1. Acemoglu, D., Laibson, D., List, J.: Microeconomics, Global Edition. Pearson Education Limited, London (2017)
2. Pindyck, R., Rubinfeld, D.: Microeconomics, Global Edition, 9th edn. Pearson, London (2017)
3. Vaz, C.R., Rauen, T.R.S., Lezana, Á.G.R.: Sustainability and innovation in the automotive sector: a structured content analysis. Sustainability. **9**(6), 1–4 (2017)
4. Wynn-Williams, M.S.: Economies of scale in the automobile industry. In: Surfing the Global Tide. Palgrave Macmillan, London (2009)
5. Gierszewska, G., Romanowska, M.: Analiza strategiczna przedsiębiorstwa. Polskie Wydawnictwo Ekonomiczne (2017)
6. Jaber, M.Y.: Learning Curves Theory, Models, and Applications. CRC Press, Boca Raton (2011)
7. Wyszomirski, O.: Transport miejski. Ekonomika i organizacja. Wydawnictwo Uniwersytetu Gdańskiego (2008)
8. Local Data Bank: [Online]. www.bdl.stat.gov.pl (2019). Accessed 2 Jan 2019
9. IGKM: Komunikacja miejska w liczbach. [Online]. www.igkm.pl/statystyka/ (2018). Accessed 5 Jan 2019
10. Schulze, A., MacDuffie, J.P., Täube, F.A.: Introduction: knowledge generation and innovation diffusion in the global automotive industry—change and stability during turbulent times. Ind. Corp. Chang. Oxford Univ. Press. **24**(3), 603–611 (2015)
11. Husan, R.: The continuing importance of economies of scale in the automotive industry. Eur. Bus. Rev. **97**(1), 38–42 (1997)
12. Wolański, M., Wołek, M., Jagiełło, A.: Jak analizować efektywność finansową i ekonomiczną napędów alternatywnych? Biul. Komun. Miej. **148**, 6–12 (2018)
13. Ally, J., Pryor, T.: Life cycle costing of diesel, natural gas, hybrid and hydrogen fuel cell bus systems: an Australian case study. Energy Policy. **94**, 285–294 (2016)
14. Nylund, N.-O., Koponen, K.: Fuel and Technology Alternatives for Buses. Overall Energy Efficiency and Emission Performance. VTT Technical Research Centre of Finland, Espoo (2012)
15. Colander, D.: Macroeconomics, 10th edn. McGraw-Hill Education, New York (2016)
16. Perloff, J.M.: Microeconomics: Theory and Applications with Calculus, 4th edn. Pearson, London (2017)
17. Argote, L., Epple, D.: Learning curves in manufacturing. Science. **247**(4945), 920–924 (1990)
18. Ally, J., Pryor, T.: Life-cycle assessment of diesel, natural gas and hydrogen fuel cell bus transportation systems. J. Power Sources. **170**, 401–411 (2007)
19. Szumska, E., Sendek-Matysiak, E., Pawełczyk, M.: Life cycle cost assessment of urban buses equipped with conventional and alternative propulsion drive. Pr. Nauk. Politech. Warsz. Transp. **120**, 395–404 (2018)
20. Clark, N.N., Zhen, F., Wayne, W.S., Lyons, D.W.: Transit Bus Life Cycle Cost and Year 2007 Emissions Estimation. [Online]. www.trid.trb.org/view/814667 (2007). Accessed 3 Jan 2019

Railway Special Grid in Near Field Communication Technology for Rail Transport Infrastructure

Arkadiusz Kampczyk

Abstract The paper presents the results of research on the author's solution employed in surveying Railway Special Grid (KOS) signs, based on the Near Field Communication technology (NFC) and reflective (measuring) target. The surveying Railway Special Grid points are at the same time track axis adjustment signs, included in the Railway Surveying Grid (KOG) and having a character of multifunctional surveying signs. The innovative surveying Railway Special Grid sign is equipped with an NFC tag, allowing wireless exchange of information (read and write). It is coated with reflective surveying foil (reflective measuring target) with crosshairs (horizontally and vertically). It determines the *X*, *Y*, *H* coordinates and track axis adjustment data. Thus, it allows measurement with an electronic tachymeter (Electronical Total Station) in relation to the surveying Railway Special Grid points. Black crosshairs—clearly visible on the surface of the reflective target—improve localisation possibilities and make it easier to aim for the central point, as the horizontal and vertical lines guarantee some offset to the track axis, while the NFC tag provides wireless exchange of data. The paper presents the author's observations and conclusions and was made as a part of statutory research 16.16.150.545.

Keywords Railway Special Grid · KOS · Railway Surveying Grid · KOG · Near Field Communication · NFC · Railway infrastructure · Reflective (measuring) targets · Reflective survey targets · Tag

---

A. Kampczyk (✉)
Faculty of Mining Surveying and Environmental Engineering, Department of Engineering Surveying and Civil Engineering, AGH University of Science and Technology, Krakow, Poland
e-mail: kampczyk@agh.edu.pl

M. Suchanek (ed.), *Transport Development Challenges in the 21st Century*, Springer Proceedings in Business and Economics,
https://doi.org/10.1007/978-3-030-50010-8_8

1 Introduction

A basis for performance of surveying and diagnostic works within the railway infrastructure is marks of the surveying Railway Special Grid (KOS). The fixed marks are situated near the track, determining its location within a vertical and horizontal plane (both profile and horizontal alignment). Railway track repairs (renovations) are intended to reinstate in whole or in part the primary performance parameters of the superstructure. We can distinguish between current, main, emergency or intervention repairs. Each of these repairs requires reference to the arks of the surveying Railway Special Grid.

The authors of the works [1–7] refer to the problems of monitoring the condition of the railway infrastructure. It was stated in the paper [1] that accurate measurement of railway track geometry bears some fundamental importance for track condition quality, both at the stage of construction and during systematic repair phases. Particular attention was paid to the modular measurement system, namely Track Geometry Measuring Trolleys (TGMTs). Reference was made to the railway geodetic control network, where during the free stationing process, bearings and distances of the total station to 5–8 known back-sight points of the railway geodetic control surrounding the stationing site are measured. It was discovered in the paper [2] that traditional maintenance of tracks is an engineering decision. Actions related to maintenance are based on reference to thresholds, where each parameter resulting from the performed measurements is compared with the agreed threshold. Reference was made to track geometry parameters in the British railway industry and to the monitoring of their condition. In DB AG (de. Deutsche Bahn AG), ÖBB (de-at. Österreichische Bundesbahnen), SBB (de-ch. Schweizerische Bundesbahnen) and other countries' railways, the tracks are referenced to the marks of the surveying Railway Special Grid, stabilised on both sides of the tracks in a form of measurement points [3]. In turn, the paper [4] draws attention to the methods of track examination in relation to the control track network and presents a new method for identification of track irregularities, determined as the Regularity for Processing Sectional Measurement Data (RPSMD). The use of surveying Railway Special Grid characters has been shown in the works [8, 9].

The authors in [5] refer to numerous issues related to the application of integrated inertial technologies for dynamic diagnostics of railway tracks, including the dynamic interaction between the rolling stock and the railway road. The railway system is undergoing technological transformation, which concerns not only high-speed trains but also infrastructure [10]. Track geometry quality is a critical factor for the safety of modern railway systems. It is achieved through a number of track surveying parameters that, in essence, characterise the quality of the absolute and relative rail locations [11]. Another important issue is the stability of the rail transport system, which is particularly important in the paper [12]. The accuracy level is needed to address certain surveying and civil engineering issues [13, 14]. In turn, [15] states that GNSS (Global Navigation Satellite Systems) measurement techniques have recently taken on a major role in civil engineering and other technical fields. In [16], it is stated that the quality of the performance of the security aspects depends on the surveying monitoring. Without proper correctness of their execution and data acquisition, it is difficult to determine the safety of technical infrastructure. On the other hand, the investment process in the case of the construction objects covers a wide range of works, including surveying works and cartographic ones [17].

The paper [6] presents a new approach to continuous monitoring of railway track on long sections. This concept makes use of fibre optics attached to rails, monitoring the emerging deformations. While verifying the fibre optic measurements, some surveying examinations were carried out in two stages. Field measurements adopted the Leica GMP111 miniprism, which was placed in a special hole of a rail clamp. While in the second phase some reflective film (reflective measuring target, reflective survey target) was attached to the clamps, thus carrying out the contactless measurements. The reflective survey target has also been used in surveying works in the expertise of constructions condition [18, 19].

The authors in [7] present a survey control network for a high-speed railway. In order to maintain high speeds for rolling stock, the condition of the track superstructure should be maintained with high precision. Precision in the surveying of railway lines is setting up horizontal and elevation control network at all levels.

The application of Near Field Communication (NFC) technology within the surveying grid has not been applied yet, especially within the Railway Surveying Grid. There is no applicable literature on NFC technology application in surveying marks, including the measurement carried out with the reflective (measuring) target. In the work [20], the author applied NFC technologies in surveying and diagnostic works applicable to the Railway Surveying Grid (KOG) and the surveying Railway Special Grid, but without the reflective (measuring) target. The current work is a continuation of the research on the application of NFC technology, called NFC + reflective (measuring) target, as the surveying Railway Special Grid.

The measurement innovation in the railway infrastructure safety is the manner of both thinking and acting, including selection of measurement techniques and methods, as well as development and utilisation of a measuring device that ensures the good quality of the acquired data [20, 21]. Innovation is one of the elements that allows to achieve a competitive advantage [22].

The authors of [23] claimed that the use of mobile devices such as smartphones and tablets has increased in recent years. As the number of mobile devices increases, more and more devices support Near Field Communication technology, which is a short-range and wireless data transmission technology. The application of mobile devices such as smartphones and tablets is found in the monitoring of rail transport infrastructure (the geometric quality of the track, traffic safety, vibration-based ride comfort, operating costs), road transport infrastructure (road surface quality, identifying braking events, the planning of road networks, road condition, operating costs, traffic safety) and travel survey methods [24–30].

In turn, the paper [31] referred to RFID (Radio-Frequency IDentification), noting that wireless communication technologies, in this case RFID, provide huge performance benefits where objects need to be identified automatically. Additionally,

RFID technology is used in the railway industry (for the position and location of trains, operation and maintenance, train axle temperature measurement, track inspection system, wagon monitoring system, in inventorying of the infrastructure of the rail transport, etc.), the automotive industry and in the Ambient Assisted Living (AAL) [32–36].

Wireless and radio technologies include:

- Bluetooth wireless technology
- Wi-Fi technology
- ZigBee wireless technology
- IrDA (Infrared Data Association)
- RFID
- Contactless smart cards
- NFC

The following hypotheses were verified in the study:

1. Potential discrepancies regarding the type of the applied surveying Railway Special Grid points.
2. The possibility to integrate the NFC technology with surveying reflective target and crosshairs as measurement points of the surveying Railway Special Grid.
3. The possibility to carry out a comparative analysis of offsets in the dy vertical plane and dx horizontal plane to the track axis, comparing the existing and designed states in real time and with the required parameters, by applying new points of the surveying Railway Special Grid KOS = NFC + reflective (measuring) target, which will guarantee the recording and reading of characteristic data of surveying monitoring of the track and turnout geometry.
4. The possibility to optimise and the universality of application of the surveying Railway Special Grid points by adopting the NFC technology and the surveying reflective target as the KOS = NFC + reflective (measuring) target.
5. The possibility to use the intersection point of straight lines of crosshairs as X, Y, H coordinates.

The main part of this paper is an innovative approach to surveying marks applied within the surveying Railway Special Grid, equipped with NFC technology and some reflective (measuring) target, called the KOS = NFC + reflective (measuring) target. Control measurements were carried out on two railway lines (railway line No. 143 "Kalety—Wrocław Popowice WP2"—section A, and railway line No. 144 "Tarnowskie Góry—Opole Główne"—section B), with a primary category within the territory of Poland. The paper presents the author's solutions, observations and conclusions. This paper was made as a part of statutory research 16.16.150.545.

2 Railway Special Grid

The surveying Railway Special Grid constitutes a collection of points including track axis adjustment marks stabilised on the traction poles, for which the grid and elevation coordinates within the assumed coordinate system were determined mathematically. The surveying Railway Special Grid is included in the Railway Surveying Grid.

KOG is a structured set of surveying points, for which their relative positions and accuracy of location are defined mathematically; KOG points are a reference for all surveying works on railway areas. The track axis adjustment marks are embedded in a permanent and stable manner on a traction pole, determining the track location on the horizontal and vertical plane.

The performed examinations proved some discrepancies in the type of adopted KOS marks. The previously adopted KOS marks are of a pin type (Fig. 1a), box type—made from aluminium and fixed to traction poles (Fig. 1b), of a painted-triangle type (Fig. 1c), and of a T-section (Fig. 1d). The KOS points which pose a part of the detailed vertical grid are at the same time the track adjustment marks.

The principles of embedding the track axis adjustment marks as KOS signs in the form of a pin type on the supporting structures (poles) of the traction network are specified in the guidelines Ig-6 [37] (Fig. 1a) and the technical standard "About the organisation and execution of measurements in railway surveying GK-1" [38]. The principles and accuracy of surveying measurements for the establishment of multi-functional marks of track axis regulation are specified in the technical standard Ig-7 [39]. The KOS mark of the box type (Fig. 1b) results from the non-binding instruction about on organisation and execution of measurements in railway surveying D-19 [40].

The task of the Railway Surveying Grid located within the railway routes and stations is to carry out not only implementation works, but also inventory works and ensure the operation of railway lines and stations. The fundamental element of conducting surveying works involving the operation of high-performance machines, especially tamping machines, are surveying marks KOS and—developed on their basis—track axis adjustment protocols.

The adjustment of the track axis is carried out in reference to the KOS characters, in relation to the profile (in the vertical plane) by raising the track to the designed ordinates. On the other hand, in the line (in the horizontal plane) by slipping the track to the axis specified in the design.

Precision requirements for the operation work and acceptance work to be carried out and the measurements to be made within it depend on speed.

These measurements are carried out in relation to the KOS characters. The values of permissible deviations of track position in relation to the KOS characters are specified by the Id-1 (D-1) [41] regulations as differences:

- Horizontal—from the KOS signs to the track axis
- Vertical—from the KOS marks to the upper rolling surface of the rail head (crown of the rail)

Fig. 1 Present marks of the surveying Railway Special Grid: (**a**) pin type (a railway line No. 661 "Dąbrowa Górnicza Towarowa DTA R5—Kozioł"), (**b**) box type (a railway line No. 143 "Kalety—Wrocław Popowice WP2"), (**c**) painted-triangle type (a railway line No. 152 "Paczyna—Lubliniec"), (**d**) with a T-section (a railway line No. 171 "Dąbrowa Górnicza Towarowa—Panewniki")

in relation to the design value.

The Railway Surveying Grid determines the correct position of the track axis and the height of the rail head at a given location.

3 NFC Technology Adopted in the Railway Special Grid Marks

Near Field Communication is a contactless and wireless communication technology which radio waves in order to spread and send some information in a digital form. It is a short-range, high-frequency and radio communication standard. NFC is equipped with the Secure Element (SE) module as a form of protection. NFC is intended for short communication between devices located close to each other. Basic NFC operation modes are:

- Peer to peer
- Read-write
- Card emulation

Communication is exercised through magnetic induction of antenna loops. It is carried out on available and unlicensed radio frequencies (13.56 MHz) of the ISM band, which was initially intended for industrial, scientific and medical purposes. This communication distinguishes two types of NFC operation:

- *Passive mode*, where the initiating device (e.g. a smartphone/tablet) generates an electromagnetic field which simultaneously supplies the target device (e.g. a tag).
- *Active mode* (semi-passive), when both devices (initiating and target) are capable of generating a signal (e.g. two smartphones/tablets). When one of the devices awaits data, its electromagnetic field if turned off (it is switched into passive mode).

NFC tags are small chips which the devices can connect to. The innovative Railway Special Grid mark called the KOS = NFC + reflective (measuring) target adopts NFC PVC NTAG213 tags with a diameter of 25 mm (Fig. 2). They are made from PVC and are water-resistant. They allow wireless exchange of information (write and read), and it is enough to hold a smartphone/tablet close to the NFC tag in order to initiate communication. It is compatible with the following standards:

- ISO 14443A
- NDEF
- NFC Forum Type 2 Tag

The employed tags have the NTAG213 chip integrated, which ensures compatibility with all popular NFC devices with an active NFC module. The NFC tags operate at an RFID frequency level of 13.56 MHz; they have 188 B memory (144 B devoted to a user). NFC tags can be secured against re-writing or removal of data.

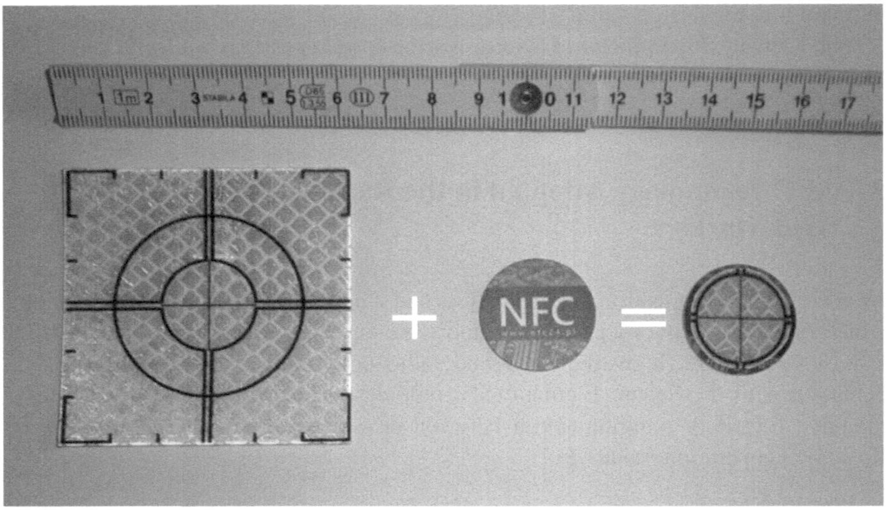

Fig. 2 Innovative mark of the surveying Railway Special Grid—components of the mark KOS = NFC + reflective (measuring) target

Table 1 NFC PVC NTAG213 characteristics

Type of characteristics	Description
Tag type: ISO 14443-3A	NXP MIFARE Ultralight (Ultralight C) NTAG213
Available technologies	NfcA, Mifare Ultralight, Ndef
Serial number	04:D5:E8:52:A3:40:80
ATQA	0x0044
SAK	0x00
Data format	NFC Forum Type 2
Capacity	144 bytes
Writeable	Yes
Possibility to write just in a read-only mode	Yes

Any information can be written or read wirelessly from the NFC tags. They have no external power source, thus they do not require any exchange or charging of a battery. The characteristics of NFC PVC NTAG213 are presented in Table 1.

The NFC PVC NTAG213 tag was fitted with geodetic surveying reflective film (reflective measuring target), the parameters of which allow the laser beam to be reflected and the measurement works to be carried out with tachymeters. A well-visible black cross is put on the film surface, facilitating localisation and targeting its centre point on a silver surface. The designed KOS = NFC + reflective (measuring) target formula with distinct horizontal and vertical lines which are just 0.5 mm thick, ensures precise pointing to the centre (Fig. 2). The employed KOS = NFC + reflective (measuring) target mark was adopted along with the electronic tachymeter TC407 Leica No. 697413 (Fig. 3). The author's marks of the surveying Railway Special Grid KOS = NFC + reflective (measuring) target were

Fig. 3 Measurement with the KOS = NFC + reflective (measuring) target

installed on traction poles on 01.06.2017 (railway line No. 143) and on 02.03.2018 (railway line No. 144). They ensure wireless exchange of information (both writing and reading).

They include data:

- On *X*, *Y*, *H* coordinates
- Offsets in the d*y* vertical plane and d*x* horizontal plane to the track axis (Fig. 3)
- km-point within the railway line
- Pole number (pole location) which is the number of the KOS=NFC + reflective (measuring) target mark
- Mark description
- Designed height of the rail head
- Existing height of the mark
- Difference in height between the designed rail head and the mark
- Designed distance of the track axis from the mark
- Designed distance between the track centres

Horizontal line—means a reference point for measurements carried out on the vertical plane (height) to the crown of the rail. Vertical line—means a reference point for measurements carried out on the horizontal plane towards the track axis (i.e. internal edge of the rail head enlarged by half of the value of the track width). In turn, the point where the straight lines cross (the centre) determines the *X*, *Y*, *H*

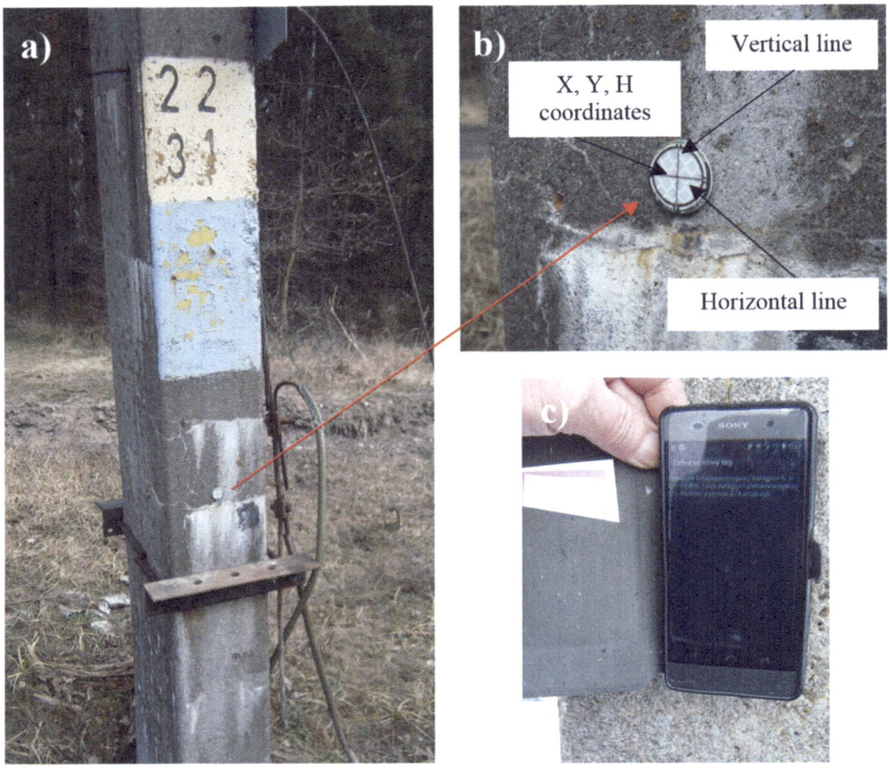

Fig. 4 KOS = NFC + reflective (measuring) target: (**a**) embedded on a traction pole of the railway line No. 144, (**b**) characteristics of the basic data, (**c**) reading via a smartphone

coordinates (Fig. 4). The KOS = NFC + reflective (measuring) target holds some data from transfer and receipt protocols of the track axis adjustment marks.

4 Conclusions

The scope of the study included an innovative introduction of the KOS = NFC + reflective target in order to improve the identification and quality of data acquisition.

The presented KOS = NFC + reflective (measuring) target mark improves the quality of surveying and diagnostic works carried out on the railway infrastructure, especially of the railway superstructure. The technology employed within the surveying Railway Special Grid marks requires much less energy, or does not require any energy at all, when compared to e.g. Bluetooth. NFC tags take energy from another source, e.g. a smartphone. Renovations of railway superstructure which requires works to maintain and reinstate the technical fitness determined with technical and exploitation parameters are referred to KOS = NFC + reflective (measuring) target marks. The presented solutions can be also introduced into monitoring of tram and metro lines infrastructure monitoring.

The innovative KOS mark equipped with an NFC tag allows wireless exchange of information (writing and reading). It determined the *X, Y, H* coordinates and data for track axis adjustment (profile and horizontal alignment). Crosshairs on the film allow measurement with an electronic tachymeter in relation to KOS marks. It facilitates localisation and targeting its centre point, and horizontal and vertical lines ensure offset of the track axis, while the NFC tag enables wireless data exchange. The researches proved the hypotheses to be correct.

The presented research results refer to the topics of contemporary studies within the diagnostics of communication construction as well as engineering and industrial surveying, with a focus on surveying in rail transport.

References

1. Chen, Q., Niu, X., Zuo, L., Zhang, T., Xiao, F., Liu, Y., Liu, J.: A railway track geometry measuring trolley system based on aided INS. Sensors. **18**(2), 538 (2018). https://doi.org/10.3390/s18020538
2. Faiz, R.B., Singh, S.: Condition monitoring of track geometry in UK Rail. In: Computing, Engineering and Information, 2009. ICC'09. International Conference on Computing, Engineering and Information. IEEE. Fullerton, CA, USA (2009). https://doi.org/10.1109/ICC.2009.9
3. Lichtberger, B.: Handbook railroad track: subgrade, permanent way, maintenance, economic efficiency. (Handbuch Gleis: Unterbau, Oberbau, Instandhaltung, Wirtschaftlichkeit). DVV Media Group GmbH/Eurailpress, Hamburg (2010)
4. Li, Y., Cen, M., Zhang, T.: A novel data processing method for sectional rail measurements to detect track irregularities in high-speed railways. Proc. Inst. Mech. Eng. Part F J. Rail Rapid Transit. **232**(2), 435–444 (2016). https://doi.org/10.1177/0954409716675002
5. Boronahin, A.M., Filatov, Y.V., Larionov, D.Y., Podgornaya, L.N., Shalymov, R.V.: Measurement system for railway track condition monitoring. In: 2015 IEEE NW Russia, Young Researchers in Electrical and Electronic Engineering Conference (EIConRusNW). St. Petersburg, Russia (2015). https://doi.org/10.1109/EIConRusNW.2015.7102252
6. Klug, F., Lackner, S., Lienhart, W.: Monitoring of railway deformations using distributed fiber optic sensors. In: Proc. Joint International Symposium on Deformation Monitoring (JISDM). Joint International Symposium on Deformation Monitoring (JISDM). Wien, Austria. https://www.tugraz.at/fileadmin/user_upload/Institute/IGMS/laboratory/FOS_Calib/2016_JISDM_Klug_et_al.pdf (2016). Accessed 21 May 2018
7. Wang, X.Y., Yang, F.Y., Gao, J.F.: Established method of high-speed railway survey control network. In: MATEC Web of Conferences 124, 01001 (2017). ICTTE (2017) https://www.matec-conferences.org/articles/matecconf/pdf/2017/38/matecconf_ictte2017_01001.pdf. https://doi.org/10.1051/matecconf/201712401001. Accessed 23 May 2018
8. Ablinger, P.: Slab track surveying and set-up—a system concept. EI Eisenbahningenieur. **52**(9), 15–21 (2001)
9. Iliffe, J.C., Arthur, J.V., Preston, C.: The snake projection: a customised grid for rail projects. Survey Rev. **39**(304), 90–99 (2007). https://doi.org/10.1179/003962607X165041
10. Gawronek, P., Makuch, M., Mitka, B., Gargula, T.: Measurements of the vertical displacements of a railway bridge using TLS technology in the context of the upgrade of the polish railway transport. Sensors. **19**(19), 1–19 (2019). https://doi.org/10.3390/s19194275

11. Gikas, V.: Geodetic survey and track recording vehicle data combined for the accurate setting out of rails on slab track. J. Spat. Sci. **50**(2), 13–23 (2010). https://doi.org/10.1080/14498596.2005.9635046
12. Berbey, A., Galán, R., San Segundo, P., Sanz Bobi, J.D., Caballero, R.: A real time rescheduling algorithm based in Lyapunov stability's index for metro lines. Un algoritmo de replanificación en tiempo real basado en un índice de estabilidad de Lyapunov para líneas de metro. Revista Iberoamericana de Automática e Informática Industrial RIAI. **11**(2), 167–178 (2014). https://doi.org/10.1016/j.riai.2014.03.005
13. Maciuk, K.: Advantages of combined GNSS processing involving a limited number of visible satellites. Sci. J. Silesian Univ. Technol. Ser. Transp. **98**, 89–100 (2018). https://doi.org/10.20858/sjsutst.2018.98.9
14. Maciuk, K.: Monitoring of Galileo on-board oscillators variations, disturbances & noises. Measurement. **147**, 1–7 (2019). https://doi.org/10.1016/j.measurement.2019.07.071
15. Maciuk, K.: GPS-only, GLONASS-only and combined GPS+GLONASS absolute positioning under different sky view conditions. Tehnički Vjesnik. **25**(3), 933–939 (2018). https://doi.org/10.17559/TV-20170411124329
16. Kampczyk, A., Dybeł, K.: Surveying monitoring of bridge structures constructed with the sliding method: how to interpret the documentation correctly? (Geodezyjny monitoring obiektów mostowych w technologii nasuwania podłużnego: jak prawidłowo interpretować dokumentację?). Nowa Geodezja w Praktyce. **45**, 1–8 (2019)
17. Kampczyk, A., Dybeł, K.: About as-built survey of a structure. (O geodezyjnej inwentaryzacji powykonawczej obiektu budowlanego). Inżynieria i Budownictwo. **6**, 327–329 (2018)
18. Dybeł, K.: Surveying works in expertises of building technical condition: how to perform—basement monitoring, Pt. 1. (Prace geodezyjne w ekspertyzach stanu technicznego budowli: jak wykonywać—monitoring fundamentów, Cz. 1). Nowa Geodezja w Praktyce. **38**, 1–5 (2018)
19. Dybeł, K.: Surveying works in expertises of building technical condition: how to perform—basement monitoring, Pt. 2. (Prace geodezyjne w ekspertyzach stanu technicznego budowli: jak wykonywać—monitoring fundamentów, Cz. 2). Nowa Geodezja w Praktyce. **39**, 1–5 (2018)
20. Kampczyk, A.: Irregularities in level crossings and pedestrian crossings. Challenges of urban mobility, transport companies and systems. In: 2018 TranSopot conference. Springer Proceedings in Business and Economics, pp. 249–261. Springer, Cham (2019). https://doi.org/10.1007/978-3-030-17743-0_21
21. Kampczyk, A.: Measurement innovations in railway infrastructure safety. World Sci. News. **89**, 336–347 (2017)
22. Wojtaszek, H., Miciuła, I.: Analysis of factors giving the opportunity for implementation of innovations on the example of manufacturing enterprises in the Silesian Province. Sustainability. **11**(20), 1–28 (2019). https://doi.org/10.3390/su11205850
23. Pesonen, J., Horster, E.: Near field communication technology in tourism. Tour. Manag. Perspect. **4**, 11–18 (2012). https://doi.org/10.1016/j.tmp.2012.04.001
24. Paixão, A., Fortunato, E., Calçada, R.: Smartphone's sensing capabilities for on-board railway track monitoring: structural performance and geometrical degradation assessment. Adv. Civil Eng. **2019**, (2019). https://doi.org/10.1155/2019/1729153
25. Mednis, A., Strazdins, G., Zviedris, R., Kanonirs, G., Selavo, L.: Real time pothole detection using Android smartphones with accelerometers. In: 2011 International Conference on Distributed Computing in Sensor Systems and Workshops (DCOSS). Barcelona, Spain, pp. 1–6 (2011) https://doi.org/10.1109/DCOSS.2011.5982206
26. Bhoraskar, R., Vankadhara, N., Raman, B., Kulkarni, P.: Wolverine: Traffic and road condition estimation using smartphone sensors. In: 2012 Fourth International Conference on Communication Systems and Networks (COMSNETS 2012). Bangalore, India, pp. 1–6 (2012). https://doi.org/10.1109/COMSNETS.2012.6151382
27. Higuera de Frutos, S., Castro, M.: Using smartphones as a very low-cost tool for road inventories. Transp. Res. Part C Emerg. Technol. **38**, 136–145 (2014). https://doi.org/10.1016/j.trc.2013.11.012

28. Nitsche, P., Widhalm, P., Breuss, S., Brändle, N., Maurer, P.: Supporting large-scale travel surveys with smartphones—a practical approach. Transp. Res. Part C Emerg. Technol. **43**(2), 212–221 (2014). https://doi.org/10.1016/j.trc.2013.11.005
29. Forslöf, L., Jones, H.: Roadroid: continuous road condition monitoring with smart phones. J. Civil Eng. Architect. **9**, 485–496 (2015). https://doi.org/10.17265/1934-7359/2015.04.012
30. Azzoug, A., Kaewunruen, S.: RideComfort: a development of crowdsourcing smartphones in measuring train ride quality. Front. Built Environ. **3**(3), (2017). https://doi.org/10.3389/fbuil.2017.00003
31. Feldhofer, M., Dominikus, S., Wolkerstorfer, J.: Strong authentication for RFID systems using the AES algorithm. In: Cryptographic Hardware and Embedded Systems—CHES 2004. 6th International Workshop Cambridge, MA, USA, Proceedings, pp. 357–370. Springer (2004). https://doi.org/10.1007/978-3-540-28632-5_26
32. Malakar, B., Roy, B. K.: Survey of RFID applications in railway industry. IEEE Xplore Digital Library. In: 2014 First International Conference on Automation, Control, Energy and Systems (ACES). Hooghy, India, pp. 1–6 (2014) https://doi.org/10.1109/ACES.2014.6807999
33. Lee, L.T., Tsang, K.F.: An active RFID system for railway vehicle identification and positioning. IEEE Xplore Digital Library. In: 2008 International Conference on Railway Engineering—Challenges for Railway Transportation in Information Age. Hong Kong, China, pp. 1–4 (2008)
34. Pérez, J., Seco, F., Milanés, V., Jiménez, A., Diaz, J.C., De Pedro, T.: An RFID-based intelligent vehicle speed controller using active traffic signals. Sensors. **10**(6), 5872–5887 (2010). https://doi.org/10.3390/s100605872
35. Berbey Alvárez, A., Castillo, G., Rodriges, J.J.P.C., López Coronado, M., Martinez Pérez, B., De la Torre Diez, I., Herreros González, J.: Uses of RFID and NFC technologies in AAL communications for health: a literature review. In: 16th LACCEI International Multi-Conference for Engineering, Education, and Technology: "Innovation in Education and Inclusion", Lima, Peru (2018). https://doi.org/10.18687/LACCEI2018.1.1.9
36. Kampczyk, A., Strach, M.: Idea of applying the RFID technique in inventorying of the infrastructure of the rail transport. In: Conference: Modern Technologies and Management Systems for Rail Transport. Vol. Research and Technical Papers of Association for Transportation Engineers in Krakow. A series: Conference materials, **3**, pp. 157–166, Krakow (2013). (Koncepcja zastosowania techniki RFID w inwentaryzacji elementów infrastruktury transportu szynowego. Konferencja: Nowoczesne technologie i systemy zarządzania w transporcie szynowym). Vol. Zeszyty Naukowo-Techniczne Stowarzyszenia Inżynierów i Techników Komunikacji Rzeczpospolitej Polskiej. Oddział w Krakowie. Seria: Materiały konferencyjne, **3**, pp. 157–166, Kraków (2013)
37. Guidelines for embedding track axis adjustment marks on the supporting structures (columns) of the overhead contact line Ig-6. PKP PLK S.A. Warsaw (2011). Wytyczne dla osadzania znaków regulacji osi toru na konstrukcjach wsporczych (słupach) sieci trakcyjnej Ig-6. PKP PLK S.A. Warszawa (2011)
38. Technical standard "About organization and execution of measurements in railway surveying" GK-1, Warsaw (2015). Resolution No. 8 of the Management Board of PKP S.A. of 12 January 2016. Standard techniczny "O organizacji i wykonywaniu pomiarów w geodezji kolejowej" GK-1, Warszawa (2015). Uchwała Nr 8 Zarządu PKP S.A. z dnia 12 stycznia 2016 r
39. Technical standard specifying the principles and accuracy of surveying measurements for the establishment of multifunctional track axis adjustment marks Ig-7. PKP PLK S.A. Warsaw (2018). Standard techniczny określający zasady i dokładności pomiarów geodezyjnych dla zakładania wielofunkcyjnych znaków regulacji osi toru Ig-7. PKP PLK S.A. Warszawa (2018)
40. Instruction about on organization and execution of measurements in railway surveying D-19, Warsaw (2000). Instrukcja o organizacji i wykonywaniu pomiarów w geodezji kolejowej D-19. Warszawa (2000)
41. Technical conditions for maintaining track surface on railway lines Id-1 (D-1). PKP PLK S.A. Warsaw (2015). Warunki techniczne utrzymania nawierzchni na liniach kolejowych Id-1 (D-1). PKP PLK S.A. Warszawa (2015)

Future of Air Cargo in Poland

Elżbieta Marciszewska, Paweł Zagrajek, and Adam Hoszman

Abstract Air cargo transport constitutes an important part of air transport market and contributes significantly to the economic development. Available data, however, indicate that the Polish market is significantly lagging behind the other European countries, in particular when considering the size of the passenger air transport market and Polish economy. The aim of the paper is to identify possible barriers and reasons behind the weakness of Polish air cargo market like regulations, infrastructure, route network and others. The authors also attempt to identify possible ways to stimulate the Polish market including infrastructure development, route network expansion with particular emphasis on wide-body aircraft and freighters as well as possible cooperation between stakeholders. The authors also attempt to depict possible scenarios of development of Polish air cargo market. The research is based on literature review, analysis of the presentations and discussion held during Future of Air Cargo in Poland conference organised by Warsaw School of Economics on 21st of February 2019, individual interviews with air cargo managers and own experience and desk research.

Keywords Air transport · Air cargo · Air freight · Transport infrastructure · Air carriers · Cargo transport market

E. Marciszewska (✉) · P. Zagrajek · A. Hoszman
Institute of Infrastructure, Transport and Mobility, Warsaw School of Economics, Warsaw, Poland
e-mail: emarci@sgh.waw.pl; pzagra1@sgh.waw.pl; ahoszm@sgh.waw.pl

M. Suchanek (ed.), *Transport Development Challenges in the 21st Century*, Springer Proceedings in Business and Economics,
https://doi.org/10.1007/978-3-030-50010-8_9

1 Introduction

The purpose of the article is a holistic analysis of trends in force under key air cargo stimulators and their potential impact on the future of this sector in Poland. Parcel services, which may be treated as a separate air freight sector, have not been analysed.

The future of air cargo in Poland depends on the future shape of key stimulants, including:

- GDP development
- Geographical market location
- International trade
- Commodities type
- Airlines capacity
- Airports capacity and infrastructure

The methodology employed in this research was based on comparative and cross-sectional data analysis. Data in time series was also analysed in order to identify prevailing trends.

2 Overall Characteristics of Polish Air Freight Market

Remarkably, due to the characteristics of the Polish market (lack of airlines' and airports' capacity), a substantial portion of air freight is taken out of the country to start its air journey. Estimates show that even 85% of air cargo from Poland is trans-

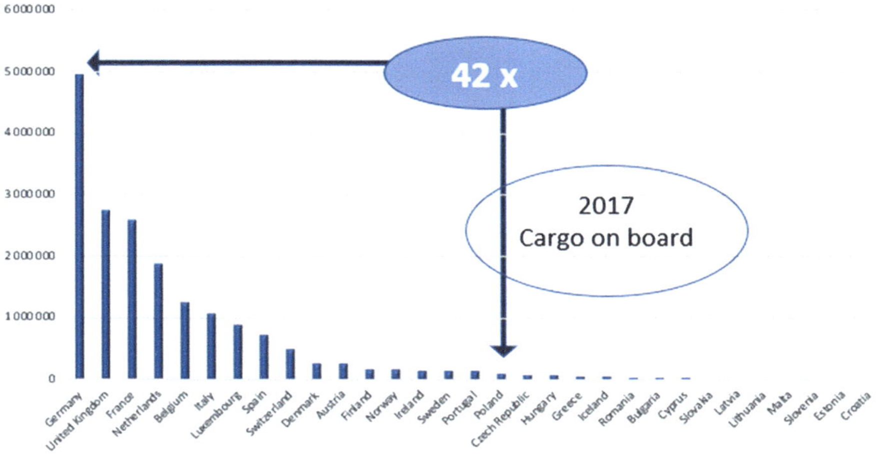

Fig. 1 Cargo on-board in the EEA countries in 2017. (Source: own compilation based on Eurostat data)

ported under the Road Feeder Service (RFS) model. Only 15% travels by air [1]. On the other hand, this should not be considered an exceptional situation as feeding cargo hubs by trucks is relatively common in Europe and North America [2]. It is estimated that air cargo shipments from Poland are shipped mainly from the main airports in Western Europe, including Frankfurt, Amsterdam, Paris, London, Vienna, and Leipzig and from airports in Central and Eastern Europe: Prague and Budapest.

One may easily expect that the above presented situation significantly impacts the position of Poland on the European cargo market. As shown by the Eurostat data illustrated on the diagram below (Fig. 1), when it comes to cargo on-board Polish market significantly lags behind some markets in Western Europe also if we consider the size of the economy and passenger traffic.

The table below (Fig. 2) shows that Chopin Airport in Warsaw, which accounts for ca. 80% of cargo on-board, has dominated this segment of the Polish market. However, the data do not fully reflect the reality of the market as RFS statistics are missing. For example, the Lodz Airport is amongst the leaders when account is taken of RFS services but as a result of the absence of a developed airline network, recently it doesn't record any cargo on-board.

Looking at the above table, we can see that the growth rates of air cargo traffic have reached two-digit numbers in recent years. As mentioned before, the data do not show the full picture of the market since RFS data is missing. In the absence of comprehensive statistics, we must use market growth estimates presented to the conference *Future of Air Cargo in Poland*. According to them, air cargo market

Airport	2012	2013	2014	2015	2016	2017
Warsaw, Chopin Airport	45 362 726	48 219 106	53 474 582	58 284 042	72 186 365	84 389 442
Katowice Pyrzowice	10 429 172	8 598 965	14 197 979	14 523 862	15 586 274	15 233 859
Gdansk Lech Walesa Airport	3 932 081	4 326 182	4 179 818	4 452 205	4 186 389	5 118 050
Kraków - Balice	583 709	465 882	0	0	3 487	108
Rzeszów Jasionka	227 076	186 678	804 703	3 863 349	731 770	392 779
Lodz Lublinek	376 266	0	0	0	0	0
Poznań Ławica	177 393	168 052	145 900	260 623	212 291	466 397
Wrocław Strachowice	95 961	93 049	115 930	89 272	2 318 334	944 055
Szczecin Goleniów	77 570	41 751	21 397	53 566	278 775	129 639
Bydgoszcz Szwederowo	25 294	1 940	1 511	7 583	0	23 545
Zielona Góra Babimost	21	0	0	0	0	0
Lublin Airport	0	0	656	14 968	1 007	14
Warsaw Modlin	0	0	0	0	0	0
Olsztyn – Mazury	0	0	0	0	0	0
Radom – Sadków	0	0	0	0	0	0
Total	61 287 269	62 101 605	72 942 476	81 549 470	95 504 692	106 697 888

Fig. 2 Air cargo in Poland in kg on-board in the years 2012–2017. (Source: own compilation based on data from the Civil Aviation Authority)

grew by as much as 30% in 2017 while in 2018 the growth was much less dynamic and reached only 3%.

3 The Forecast of Air Cargo Market Development

According to data provided by Boeing [3], long-term average annual growth rate of air cargo traffic amounted to 4.2%, whereas in 2017 it reached 10.1%. The company believes this high growth rate was driven by the global economic expansion, increasing industrial production, and global trade growth. Boeing argues that despite the drop in the growth rate in 2018[1] many indicators suggest that there are strong reasons to believe that in the future cargo will be growing above the long-term average. In their opinion, further growth of cargo will be stimulated by the economic growth and dynamic growth of e-commerce. Boeing also observes that goods, which due to their inherent features are most preferably transported by air (time-sensitive perishables, high-value commodities), are some of the fastest growing trade flows around the world. On the side of threats to the growth of air cargo, Boeing lists trade wars and volatile fuel prices. The company also expects further dynamic growth of the Asian market that is crucial for Europe. According to Boeing, intra-European traffic and cargo flows between Europe and North America will grow much below the world average. Nevertheless, having analysed all forecasted flows to and from Europe, the outlook for the development of air cargo market in Europe should be assessed as positive. Regrettably, no forecast is provided individually for Polish market.

4 GDP Growth Forecast

There is a high correlation between air freight and economic growth. Cargo market is also more responsive to general economic situation than passenger market [4]. Freight traffic however tends to fall faster in comparison to economic downturn and rebounds faster on the economic up-cycle [5]. During IATA Global Airport and Passenger Symposium [6], global economic slowdown, trade war and geopolitical tensions at various parts of the world were indicated among the major threats to the air transport development in the near future. Taking into consideration pessimistic predictions concerning major economic and political factors, it can be assumed that air cargo traffic will record negative trend in the near term both globally and locally in Europe.

This negative scenario may be realised in Poland. Current forecasts of European Commission however, although already lowered, show that the GDP will grow at

[1] Boeing estimates suggest the growth reached ca. 4%.

the rate of 3.5 and 3.2% in 2019 and 2020, respectively [7]. This could provide strong foundations for the growth of air cargo in the forthcoming couple of years, provided that Polish market will be not negatively hit by global crisis.

5 Geographical Market Location and International Trade

When examining the potential of the cargo market in Poland, we need to look at it from the viewpoint of the market location. As demonstrated by data provided by IATA WATS [8], air cargo traffic inside Europe accounts for only 2.3% of global air traffic. However, globally Europe is a very important market in terms of intercontinental transport. Two out of three of the biggest flows of goods transported by air start and end in Europe. The biggest traffic destination from Europe is Asia (19.4% of the global traffic).

Geographic structure of Polish foreign trade is an important factor that impacts the growth of air cargo market and it is directly linked to geographical location of Polish market. From the viewpoint of air cargo market and its growth, we need to bear in mind that air transport has got an advantage over other modes of transport in situations when a large portion of trade flows are targeting countries located at very long distances. At the same time, as demonstrated by data, Polish trade focuses mainly on the EU Member States. When it comes to exports, more than 30% of trade takes place between Poland and its immediate neighbours: Germany and Czech Republic. The list of top 10 importers of Polish goods includes only one country from outside of the European Union, i.e. the United States. Slightly more promising, from the point of view of air cargo growth potential, is the structure of Polish imports. China ranks second amongst the biggest exporters to Poland. The list of top 10 exporters to Poland also features the United States. Nevertheless, also in imports, most trade flows originate from the European continent.

6 Airlines Capacity

Air is the most expensive mode of transporting goods. Prices are 10–15 times greater for air transport than for sea transport [5]. That is because aircraft is costly to operate, especially per 1 kg of cargo.

As much as ca. 50% [3] of air cargo is carried in the bellies of passenger aircraft. These aircraft can hold up to 25 tons of cargo if they are wide-body aircraft and in extreme cases up to 5 tons if they are big narrow-body aircraft (usually maximum 1–2 tons). In fact, cargo capacity of a given flight depends on many factors. It should be stressed that destinations operated by passenger aircraft are selected almost exclusively based on the passengers' demand. It means, cargo transport on these routes is not driven by the needs and/or demand of the cargo industry. The biggest cargo transport possibilities are offered by passenger planes in the freight configuration. The largest freighter can carry even up to 140 tons,[2] while the capacity of a medium-sized wide-body freighter aircraft reaches ca. 50 tons. Operation-wise, such aircraft can be used in two ways: on regular cargo routes or in ad hoc operations.

Extensive network of flight routes operated by passenger wide-body aircraft or freighters is especially important for the growth of air cargo market. In this context, attention should be drawn to the fact that nowadays using wide-body aircraft for moving cargo, despite significant increases reported in recent years, remains still rare in Poland. PLL LOT operates the largest fleet of wide-body aircraft in Poland comprising currently[3] eight Boeings 787-8 and three 787-9. On top of that, four Boeings 787-9 are expected to be delivered to the air carrier in 2019. Notably, some services by wide-body airplanes are operated by PLL LOT from Budapest. On routes to and from Warsaw wide-body aircraft are operated also by Emirates, Qatar Airways, Air China, to a limited extent by Turkish Airlines, and Air Canada Rouge as a seasonal service. It is worth adding that regular routes operated by freighters to and from Poland are available only from courier service companies. The only carrier in Poland who operates a wide-body aircraft for that purpose is UPS flying Boeing 747 from Warsaw in different configurations. So far, no air carrier has decided to start operating regularly scheduled cargo flights, the so-called general cargo, from Poland. For the time being[4] not a single carrier has announced such intentions. Yet we should mention that there is a carrier in the Polish market, Sky Taxi, who operates a wide-body freighter Boeing 767. The aircraft is used exclusively on an ad hoc basis mostly outside of Poland. Thus Polish air cargo network has got an increasing albeit still limited potential stimulating the growth of air cargo market.

7 Air Cargo Potential in the Light of Growth of the Production of Certain Commodity Groups

Due to high cost of transport and operational limitations, air cargo represents less than 1% of the global trade by tonnage [8]. Yet air cargo surely outperforms other modes of transportation when it comes to delivery time, which becomes especially competitive on long-haul flights. Characteristics of air cargo industry make them best placed to carry specific groups of products, which in a simplified approach can be described by three criteria:

- Sensitive to the length of the journey
- Sensitive to temperature
- High-value goods
- Rarely unique cargo—oversized and superheavy goods [9]

---

[2] Boeing 747, version 800.

[3] 20 February 2019

[4] 1 March 2019

The above characteristics should be seen as a rather broad and generalised description of goods that are most frequently carried by air. Goods carried by aircraft may have one or more of the above-mentioned features, they may, e.g. be sensitive to the length of the journey and to temperature.

Considering the above criteria we may say that the following are mostly transported by air: mail, including diplomatic pouches, courier parcels, pharmaceuticals and other medical products, medical samples, live animals, human organs, human remains, electronics and accessories, parts for aircraft, automotive, mining, maritime, etc., industry, food (e.g. meat, fish), flowers, radioactive materials, clothes, money, valuables and artworks.[5]

Because air cargo tends to specialise to carry specific commodities, it tends to develop in a given country or region in line with the presence of particular branches of industry, such as electronics, pharmaceutical, automotive, aircraft, maritime or extraction industry. Air cargo is also stimulated by e-commerce.

Thus we may propose a thesis that the future pace of air cargo development on routes from Poland will partly depend on the expansion of the production of goods which, due to their features, are the most predestined to be transported by air.

Speaking of air cargo traffic we should take a closer look at aircraft industry. This is an industry with long tradition in Poland. Nowadays, 90% of aircraft-related production in Poland is located in Podkarpacie [11]. It is hard to estimate the rate, with which aircraft industry will continue to grow in Poland, but we may assume that it will be meaningful. As suggested by Magazyn Przemysłowy [11], the output will be driven by global orders for aircraft together with growing market in Poland and orders from the civil sector.

Pharmaceutical industry is another sector of the economy that can significantly stimulate the growth of air cargo. According to Śledziewska [12], currently pharmaceuticals account for 1.33 of the GDP of Poland, which is lower than in other EU Member States. Nevertheless, the author draws attention to the fact that the industry share in GDP grows for Poland twice as quickly as for other countries. The growth of Polish air cargo surely largely depends on the automotive industry. Car and spare parts manufacturing in Poland represents as much as 8% of Poland's GDP, generates 13% of exports and grows dynamically. Experts forecast that the turnover of the automotive sector in Poland will be increasing also in the future [13]. The situation of the Polish shipyard industry looks differently. Authors of the report titled "Polski sektor stoczniowy, stan obecny, perspektywy, zagrożenia" [*Polish Shipbuilding Industry: Current Situation, Perspectives, and Threats*] [14] indicate that shipbuilding industry in Western Europe has expanded its order portfolio in recent years. Yet, in the opinion of the authors, Polish shipyards have not benefited from these positive trends because, inter alia, they have not signed any significant long-term contracts for the construction of special-purpose vessels. According to the report, Polish shipyards are not profitable and margins that they achieve do not suffice to provide a stable capital surplus that would secure their further growth. When examining the

[5] For a broader scope of goods carried by air, see: J. G. Wensveen [10].

growth potential of Polish air cargo traffic through the lenses of the development of certain industries, which can boost this growth, we should conclude that the picture is not clear. However, if we consider positive outlook for the growth of the pharmaceutical, automotive and aircraft industries we may argue that industrial output will stimulate increases in air cargo traffic in Poland.

8 Air Cargo Infrastructure in Poland

Infrastructure used for cargo handling is also an important component of the landscape that conditions the growth of air cargo traffic. As shown by the debate held during the conference *Future of Air Cargo in Poland* at the Warsaw School of Economics [15] [6] the growth of air cargo infrastructure in Poland can be perceived at multiple levels. Firstly, we need to consider the availability and capacity of cargo service terminals in the Polish market. Cargo service infrastructure is available at airports in Warsaw (Chopin Airport), Rzeszow, Krakow, Katowice, Lodz, Wrocław, Poznan, and Gdańsk. On top of that, we should also mention the terminal at the Szczecin Airport currently under construction. In Warsaw and Poznan agglomeration, there are also air cargo service terminals located outside of the airports. Participants to the debate at the conference *Future of Air Cargo in Poland* decided that the quality of services rendered by the existing terminals is not satisfactory. Interestingly, the discussion addressed the future Central Communication Port, which in the opinion of conference participants should include special infrastructure dedicated for air cargo handling.

In addition, conference participants highlighted the importance of awarding the security status of air freight for the development of air cargo infrastructure in Poland. It links with the need to protect the aviation against any unlawful interference. In the debate, attention was also drawn to the problems of awarding such status in Poland. One of the shortcomings of the Polish market is also the absence of properly trained dogs. Another element tackled in the discussion was the insufficient population of companies with the so-called "known consignor" status.[7] Participants of the conference stressed that a safe supply chain from a known consignor simplifies, shortens and facilitates the logistics process. As shown by data, Poland is poorly developed in this area compared to other countries in Europe. While there are only a dozen or so companies with this status in the Polish market, in Germany there are ca. 2k companies with the "known consignor" status.

[6] Conference titled *Future of Air Cargo in Poland*, Warsaw School of Economics, 21 February 2019.

[7] We need to stress that an air freight shipment from the so-called "known consignor" within a safe supply chain does not go through security checks at all. Simply speaking, a known consignor is a manufacturer who meets specific legal requirements, as a result of which work organisation, procedures and staff working on the dispatch of air freight guarantee that no unlawful interference can take place. For more on the subject see: http://www.ulc.gov.pl/pl/ochrona/cargo/znany-nadawca.

9 Summary and Conclusions

When examining the outlook for the growth of the air cargo market in Poland, we need to remember that Poland is a part of Europe, where perspectives for the development of air cargo traffic are good largely because of cargo flows from the Far East. Broadly understood economic growth and the development of industries whose products are transported by air are factors that surely favour the growth of air cargo in Poland. The fact that nowadays substantial flows of air cargo are carried on trucks to airports located west and south of Poland shows huge potential of the domestic market, which can increase cargo flows from Polish airports and the handling of air freight in Polish terminals not only through the organic growth of the market but mainly by capturing flows already generated by the market. Presented analyses inform that in order to achieve this, Polish cargo infrastructure should be improved in terms of terminal capacity as well as the quality and smoothness of security checks. The construction of the Central Communication Port can potentially be positive for the growth of cargo market. Another favourable factor for air cargo growth can come from the expansion of route networks for freighters or wide-body passenger aircraft. New purchases for the PLL LOT fleet in 2019 should additionally support the trend. Another stimulus for the growth of air cargo traffic in Poland can also be provided by the increasing number of "known consignor" status companies. Summing up, Polish air cargo market has got a substantial growth potential, higher than the potential of passenger traffic market. Yet the market needs changes at different levels. Interesting conclusions have been presented by the participants to the conference titled *Future of Air Cargo in Poland*, who highlighted the need for enhanced cooperation between companies in the market and a new approach to such collaboration.

References

1. Rześny Cieplińska, J., Wach-Kloskowska, M.: RFS as an Unconventional Method of Cargo Shipment by Air Transport. WSB University in Gdansk, Gdańsk (2017)
2. Morrel, P.S.: The air cargo industry. In: O'Connell, J.F., Williams, G. (eds.) Air Transport in the 21st Century. Key Strategic Developments, p. 241. Ashgate, Aldershot (2014)
3. Boeing: World Air Cargo Forecast 2018–2037 (2018)
4. Doganis, R.: Flying Off Course, 4th edn, p. 287. Routledge, London (2010)
5. Morrell, P.S.: Moving Boxes by Air. The Economics of International Air Cargo, p. 24. Ashgate, Aldershot (2011)
6. IATA Global Airport and Passenger Symposium (GAPS): What does the future hold—and economic and passenger outlook. 15–17 October 2019 Warsaw (2019)
7. TVN24bis.: https://tvn24bis.pl/z-kraju,74/ke-obnizyla-prognozy-pkb-dla-polski-na-lata-2019-2020,907442.html. Accessed 10 Mar 2019
8. IATA: WATS industry infographic. https://www.boeing.com/resources/boeingdotcom/commercial/about-our-market/cargo-market-detail-wacf/download-report/assets/pdfs/2018_WACF.pdf (2018)
9. Sales, M.: Air Cargo Management, p. 141. Routledge, London (2016)

10. Wensveen, J.G.: Air Transportation. A Management Perspective, 6th edn, pp. 329–331. Ashgate, Aldershot (2017)
11. Magazyn Przemysłowy: https://www.magazynprzemyslowy.pl/zarzadzanie-i-rynek/Polski-przemysl-lotniczy,11423,1 (2019). Accessed 3 Mar 2019
12. Śledziewska, K.: https://www.pb.pl/farmacja-jest-jednym-z-filarow-gospodarki-908794 (2018). Accessed 3 Mar 2019
13. Kaczmarczyk, M.: https://www.newsweek.pl/biznes/gospodarka/przemysl-motoryzacyjny-w-polsce-jak-rozwija-sie-motoryzacja/l5d9f8n. Accessed 3 March 2019
14. The Eastern Institute: Polski sektor stoczniowy, stan obecny, perspektywy, zagrożenia. Foundation of the Institute of Eastern Studies, Warsaw (2018)
15. SGH: Conference titled Future of Air Cargo in Poland, Warsaw School of Economics, 21 Feb 2019

Conditions of Using the Bi-directional Tram in the Sustainable Urban Transport System (Case Study on the Example of Szczecin, Poland)

Oliwia Pietrzak and Krystian Pietrzak

Abstract Tram transport is an important component of the urban transport system. In addition to subway and urban railway (including metropolitan railway), it is a component of the Urban Rail Transport (URT). One of the solutions in this system is the bi-directional tram. The objective of the article was to indicate factors conditioning the possibility of implementing and using the bi-directional tram in the urban transport system, with reference to the case study of Szczecin, Poland. The qualities of this mean of transport and conditions determining its functionality were also indicated, including those underlining its potential advantages over the traditional tram system. The analysis also includes the role that the bi-directional tram plays in European cities and potential benefits of using this mean of public transport in a sustainable urban transport system. The authors of the article applied the method of critical literature analysis, the documentary method and analysis, and the synthesis method. In addition, the results of field studies carried out by the authors in selected European cities with external and participating observations were included. The research was carried out from 2018 to 2019.

Keywords Tram · Streetcar · Urban transport · Public transport · Passenger transport · Sustainable development

1 Introduction

Urban transport plays a special role in the passenger transport system. In particular, it is the consequence of the role that the city plays in the life and development of the society. Cities have several characteristics:

O. Pietrzak (✉) · K. Pietrzak
Maritime University of Szczecin, Szczecin, Poland
e-mail: o.pietrzak@am.szczecin.pl; k.pietrzak@am.szczecin.pl

M. Suchanek (ed.), *Transport Development Challenges in the 21st Century*, Springer Proceedings in Business and Economics,
https://doi.org/10.1007/978-3-030-50010-8_10

- Limited space.
- They constitute the living environment of an increasing number of the world population [1].
- They are a place of concentration of various forms of human activity and, hence, of transport needs.
- The so-called urban lifestyle is becoming more popular and the mobility of the society is increasing.

These characteristics directly or indirectly affect the possibilities for and directions of designing and functioning of the urban transport system. City dwellers need the public transport system to support their daily activities [2].

Progressing urbanization processes are beneficial, for instance, they encourage economic growth, increase the number of jobs, raise living standards, and facilitate access to cultural, scientific, and health centers; however, they also cause numerous undesirable phenomena. They include, among others:

- Expansive development of the transport infrastructure in cities
- Increased environmental pollution and noise
- Congestion on roads
- Decreased safety of pedestrians
- Increased number of road accidents
- Increased transport costs in cities
- Increased land prices for residential construction in urban areas

The basic task of the urban transport system is to meet the transport needs of its users [3]. These needs can be meet individual or public transport. In the face of activities undertaken by cities and metropolitan centers in order to limit individual motorization, it is worth noting that public transport, apart from the indicated basic function, also performs other additional functions that are important for the city and society. The crucial functions are of social nature, for instance, handling the needs of residents who do not have access to individual vehicles, with low economic opportunities [4], residents with movement limitations, children and adolescents, people staying in the area occasionally and responding to the aging process of the society. Environmental functions are also of importance—they include counteracting road congestion and degradation of the natural environment, as well as spatial functions, such as deciding on the space of the city and its transport system.

It is also worth emphasizing the unfavorable phenomenon of "temporal remoteness" of individual areas of urban space. As a result of the growing mobility of the society, the increase in the degree of individual motorization and the progressing phenomenon of congestion on the roads, the time needed to travel a certain distance is much longer. As a result, even though the distance is clear and there are investments in the transport infrastructure or organization of the transport system, the temporary availability of specific urban areas is significantly reduced. Consequently, transport availability, determined by four main components: spatial, transport, time, and individual [5], and the attractiveness of the area are reduced. Therefore, urban transport has become a tool of the transport policy and also of spatial, economic, and social policies.

It is generally recognized that the carrot and stick approach is necessary to decrease car usage and to increase public transport usage [6]. Therefore, municipal authorities decide to implement a number of solutions which are to discourage residents from using private vehicles to travel into urban areas and indicate the possibilities and benefits of using public transport.

The objective of the article was to indicate factors conditioning the possibility of implementing and using the bi-directional tram in the urban transport system, with reference to the case study of Szczecin, Poland.

An important part of the research was to answer the following questions:

- What are the features of the bi-directional tram as a mean of public transport?
- What functions does the bi-directional tram perform in European cities?
- Is it possible to implement the bi-directional tram in Szczecin?
- What benefits does the implementation of the bi-directional tram into the urban transport system offer Szczecin?

The article presents the results of the two-stage research:

- Field studies carried out in selected European cities: Brussels (Belgium), Prague (Czech Republic), Warsaw (Poland), Cracow (Poland), Wroclaw (Poland), Olsztyn (Poland), 2018
- A case study on the example of Szczecin and the analysis of the conditions for the implementation of the bi-directional tram in Szczecin, 2019

2 Characteristics of the Tram as a Mean of Public Transport

Systems of public transport are becoming increasingly important as they are reliable, accessible, environmentally friendly and satisfy the capacity needs. These benefits help to revive tram systems in many regions of the world [7].

Tram is a passenger or freight road vehicle designed for use on a tramline [8]. According to the Regulation (EU) 2018/643, tram means a passenger road vehicle designed to seat more than nine persons (including the driver), which is connected to electric conductors or powered by diesel engine and which is rail-borne [9]. In Poland, in accordance with the Regulation of the Minister of Infrastructure, the tram is defined as a vehicle designed to carry people or goods powered by electricity, moving on rails on public roads [10].

Trams have numerous qualities which allow them to play an increasingly important role in public transport systems [7, 11]. A tram-based passenger transport system offers a more cost-efficient alternative to metro or light rail [12], and it provides higher capacity and reduced negative impact on the environment as compared to buses [13]. Modern trams are also comfortable for users and more punctual than buses [14]; they are well suited for transportation in the central city areas with high traffic [15].

The key features of the tram as a mean of urban transport are indicated in Table 1.

Table 1 Selected features of the tram compared to other means of public transport

Tram features	
Positive	Negative
The possibility of independence from traffic and congestion by separate communication routes	Hindered possibility of organizing a bypass in the event of sudden events or roadworks
Higher average and maximum communication speed compared to buses and trolleybuses [15]	Limited possibilities of current adjustments of the route
Higher capacity compared to buses and trolleybuses	Relatively high unit cost of buying a new vehicle
Reduced air pollution, noise, and vibration	Less maneuverable compared to buses
Lower investment costs compared to the metro	Higher investment costs compared to buses and trolleybuses [15]
Low sensitivity to adverse weather conditions	
Longer service life compared to buses and trolleybuses (about 40 years)	
Narrower lane compared to buses	
Easier access to passenger stops compared to the metro	
Less expensive operation compared to buses and trolleybuses (in the case of passenger traffic flow of more than 5000 passengers/hour) [15]	
Greater punctuality compared to buses and trolleybuses	
Possibility of entering narrow city roads	
Considerable transport safety	
Compliance with EU transport policy requirements	
Integration with urban space	
Positive impact on the image of the city	

A special type of tram is the bi-directional tram. The Regulation of the Polish Minister of Infrastructure defines the bi-directional tram a vehicle adapted for driving in two directions [10]. As this definition is too general, the authors decided to coin their own. They define the bi-directional tram as a vehicle adapted for driving in two directions, whose duplicate equipment (cabin, doors on both sides, pantograph) allows to perform its function regardless of the direction of movement.

The functionalities of the bi-directional tram show that such trams may fulfil all tasks of the uni-directional tram, and, in addition, its characteristic features allow it to be used in situations unavailable for traditional trams.

The crucial features of the bi-directional tram, as compared to the uni-directional tram, are indicated in Table 2. The analysis of these features allows us to state that the bi-directional tram is more universal. The use of this type of vehicle offers specific benefits to the process of handling passenger traffic in the city.

The study carried out by the authors found that the popularity of bi-directional trams is growing in urban transport systems in Europe. Despite significant differences in the spatial structure of individual cities, as well as in the system of public

Table 2 Selected features of the bi-directional tram compared to the uni-directional tram

Bi-directional tram features	
Positive	Negative
Limited land demand (no need to build a loop)	Higher total vehicle weight
Even wear of road wheels (no need to turn wagons to extend the time of their use)	Greater pressure on the track
Even and slower wear of other vehicle components, due to their duplication, e.g., equipment of the driver's cab	Higher purchase costs (the difference estimated by the manufacturers of rolling stock is around +8–15%)
Possibility of independent (without the use of a technical vehicle) exit after a collision (without the exclusion of further traffic)	Higher labor intensity of maintenance activities (e.g., periodic inspections of a larger number of doors)
Possibility of driving in a shuttle system, e.g., during roadworks or in the event of collisions	Smaller number of seats
Possibility of extending the route at any time/stage construction of the route—the route ends with a switch-back	Passengers have to get accustomed to the new solution (as compared to the traditional system)
Possibility of building the route in densely developed areas	
Lower costs and shorter route construction time (no need to build a loop)	
Possibility of quick direction changes	
Handling central and traditional platforms	
Possibility of using the existing infrastructure	
Possibility of being used as the uni-directional tram	

transport, more and more cities decide to include bi-directional vehicles in the city traffic. This mean of transport is used to a different extent, performing the basic, equivalent, or auxiliary function in the city tram system. In Europe, this type of vehicles may be found in Brussels or Prague, while in Poland it is used in Warsaw, Kraków, Wrocław, Poznań, and Olsztyn. It is worth noting that the new tram network in Olsztyn was built without tram loops—it is only served by bi-directional trams.

3 The Use of Bi-directional Trams in Selected Cities in Europe

Field studies were conducted to identify possible applications and functions of the bi-directional tram in urban transport systems. An analysis was carried out in selected European cities which use the bi-directional tram. The study was conducted in 2018. There were two groups of cities:

- Cities which have a tram network in operation from many years: Brussels (Belgium), Prague (Czech Republic), Warsaw (Poland), Cracow (Poland), Wroclaw (Poland)
- A city with a new tram network: Olsztyn (Poland)

The selected results are presented in Table 3, which contains the following data:

- Type of infrastructure: tram lines with tram loops, tram lines with switch-back terminals, temporary switch-back and switch-back between tram stops
- Type of trams: uni-directional and bi-directional
- Manner of using bi-directional trams: on the lines with loops, on the lines with switch-back terminals, and on the lines under renovation

Table 3 shows that the use of the bi-directional tram does not directly depend on the construction of special infrastructure (with switch-back terminals). Two of the analyzed cities, Prague and Cracow, only use the bi-directional tram to operate lines under renovation and in the remaining period they use it interchangeably with uni-directional trams on the lines with loops.

Table 3 The use of bi-directional trams in selected European cities

		Brussels (Belgium)	Prague (Czech Republic)	Warsaw (Poland)	Cracow (Poland)	Wroclaw (Poland)	Olsztyn (Poland)
Infrastructure	Line with tram loops	+	+	+	+	+	–
	Line with switch-back terminals	+	–	+	–	+	+
	Temporary switch-back	+	+	+	+	+	–
	Switch-back between tram stops	+	–	–	–	–	–
Tram	Uni-directional trams	–	+	+	+	+	–
	Bi-directional trams	+	+	+	+	+	+
Manner of using bi-directional trams	On the lines with loops	+	+	+	+	–	–
	On the lines with switch-back terminals	+	–	+	–	+	+
	On the lines under renovation	+	+	+	+	+	–

Warsaw and Wroclaw use the bi-directional tram to operate lines under renovation and also to operate new lines with switch-back terminals and lines constructed in stages.

The situation is different in the other analyzed cities—Olsztyn and Brussels. Olsztyn constructed an entire tram network adjusted solely to bi-directional trams (2015). The city has no tram line ending with a loop. However, many years ago, Brussels resigned from uni-directional trams, gradually adjusting its tram network to bi-directional trams as the only type of trams. However, some lines ending with a loop remained for historical reasons. Furthermore, the city network has switch-backs between tram stops, which allows the bi-directional tram to handle traffic on shortened routes.

Studies have shown that the analyzed cities use the bi-directional tram to a different extent:

- Sole mean of transport (Olsztyn, Brussels)
- Handling traffic during renovations
- Handling newly built lines with switch-back terminals
- Investment phasing

Despite the fact that the use of bi-directional tram does not depend directly on the construction of special infrastructure and it may be used interchangeably with uni-directional trams, none of the analyzed cities use bi-directional trams solely to replace uni-directional vehicles. All these cities also pursue additional possibilities offered by this mean of transport.

4 The Use of Bi-directional Trams in the Urban Transport System: Case Study on the Example of Szczecin, Poland

Szczecin is located in the north-western part of Poland. It is the capital of the West Pomeranian region and the largest urbanized center in this area. It is also the central part of the Szczecin Metropolitan Area (SMA). The area of the city is 301 km^2, thus ranking third in the country. The population of Szczecin amounted to 403,274 people as of June 30, 2018 [16].

The public transport system in the city is based on two subsystems: trams and buses. Transport services are provided under contracts concluded with one tram carrier (Tramwaje Szczecińskie Sp. z o.o.—TS) and four bus carriers. However, it is worth noting that due to the need for changes to the communication behavior of the dwellers of the city and the SMA and shifting the passenger load from individual transport to public transport, steps were taken to integrate rail transport into the transport system of the SMA by creating and implementing the Szczecin Metropolitan Railway (SMR) [17].

The assumptions of the city transport policy, included in the plan of sustainable development of public transport, indicate that the basic means of public transport in

Szczecin is and will be the tram along with the bus network. Moreover, it was emphasized that the city center will prioritize trams as means of transport, and its implementation into the public transport network is to be ensured with integrated transfer nodes [18]. Such assumptions are particularly important from the point of view of the solution proposed by the authors—the implementation of bi-directional trams in Szczecin. The solutions proposed by the authors are based on the study they carried out in 2018 for the Szczecin City Hall on technical, operational, and organizational conditions for using the bi-directional tram in the public transport system in Szczecin. It was a direct result of the need to expand the existing tram network in connection with the spatial development of the city and the purchase of vehicles to handle traffic on new routes. To this end, the city is planning to purchase new trams, therefore, there is a compelling reason to consider the use of bi-directional vehicles.

The current offer of public transport in Szczecin in the field of tram transport includes 12 day lines [19]. Traffic handled by the TS company is directed through 119 km of single tracks and 12 tram loops (two of which are street loops) [20]. The system is entirely based on uni-directional trams, including the infrastructure—all routes end with loops. It should be noted that this situation does not exclude or limit the possibility of using bi-directional trams.

Taking into consideration the research conducted in selected European cities, the analysis of the condition of the Szczecin tram system, and the necessity of further development, the authors suggested a three-step implementation process of the bi-directional tram (Table 4):

- ***First step*** (*green color in* Table 4)—the purchase of bi-directional trams and temporary switch-backs—to operate lines under renovation (similarly to Cracow and Prague)

Table 4 The use of bi-directional trams in the city of Szczecin

		Szczecin (currently)	**Szczecin (1st step)**	**Szczecin (2nd step)**	**Szczecin (3rd step)**
infrastructure	line with tram loops	+	+	+	+
	line with switch-back terminals	-	-	+	+
	temporary switch-back	-	+	+	+
	switch-back between tram stops	-	-	-	+
tram	uni-directional trams	+	+	+	+
	bi-directional trams	-	+	+	+
manner of using bi-directional trams	on the lines with loops	+	+	+	+
	on the lines with switch-back terminals	-	-	+	+
	on the lines under renovation	-	+	+	+
	on shortened routes	-	-	-	+

- ***Second step*** (*yellow color in* Table 4)—construction of new lines ending with switch-back terminals and increasing the use of the bi-directional tram to operate lines under construction (similarly to Wroclaw and Warsaw)
- ***Third step*** (*blue color in* Table 4)—the construction of switch-backs between tram stops and using the bi-directional tram on shortened routes (similarly to Brussels)

4.1 First Step: Use on the Existing Lines Under Renovation

Exploitation of transport system is associated with the need to conduct ongoing renovations of the infrastructure. The purpose of the activities which have been implemented in Szczecin is to reduce the tram traffic on the entire communication line. A tram line is usually temporarily replaced by a bus line. As a result, the replacement is included in the car traffic, thus adding to the road congestion. This solution causes bus delays and, consequently, passenger dissatisfaction. In addition, there is always a need to organize temporary bus stops, which affects the safety and comfort of passengers. The bus line cannot follow the same or similar route as the tram, at the same time keeping the timetable. These changes often disturb the integration of the network.

The use of the bi-directional tram in the indicated range can be implemented by shortening the route (Fig. 1) or temporarily leading traffic along one track of the renovated section, without shortening the starting route.

Organizers, municipal authorities, and public transport operators should always take into account the fact that **it is easier to attract new passengers than to regain the trust of those who resigned from public transport**. Therefore, renovations of the tram network have to provide the possibility of handling passengers at the current level in terms of accessibility, time, cost, convenience, and integration. The bi-directional tram offers such opportunities.

4.2 Second Step: New Tram Lines with Switch-Back Terminals

The proposed solution assumes the extension of the existing tram route (Fig. 2—*black line*) along 26 Kwietnia St. in two stages (*diagram in* Fig. 2):

- Stage I: construction of a two-track route on the section: Boh. Warszawy/Krzywoustego intersection—Witkiewicza/Derdowskiego intersection (Kaliny housing estate), approx. 2.2 km long (length of the whole line approx. 4.5 km), 6 stops, potential passengers: over 6000 residential premises, inhabited by over 15,000 residents within 400 m from the tram stop (Fig. 2—*red line*)
- Stage II: construction of a two-track route on the section: Witkiewicza/Derdowskiego intersection (Kaliny housing estate)—Zawadzkiego—Klonowica housing estate, length and mileage depending on other infrastructure investments of the city (Fig. 2—*blue line*)

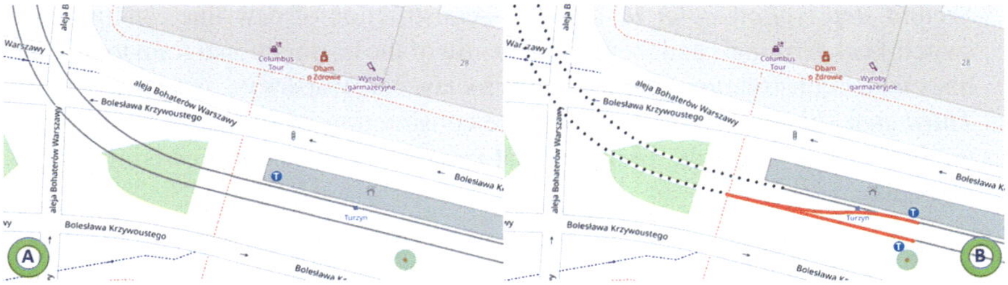

Fig. 1 Tram line in Szczecin. (**a**) Regular traffic, (**b**) A proposition of using the bi-directional tram during renovation in Szczecin. (Source: Own work based on www.openstreetmap.org)

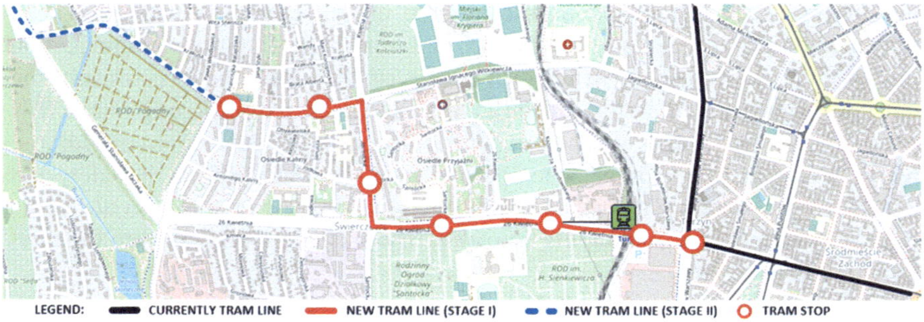

Fig. 2 The proposed new tram line using a bi-directional tram in Szczecin. (Source: Own work based on www.openstreetmap.org)

As there are some ongoing construction investment projects, it is impossible to make an intermediate loop during the entire route. The use of the uni-directional tram would force the investment to be carried out in one step. This would result in the need to incur significant costs in a short period of time, without the possibility of commencing the operation of the route at individual stages of the investment. The use of the bi-directional tram will allow for: a gradual implementation of the investment (after the construction of stage 1, the route can be completed with the switch-back) and the inclusion of trams at the first stage of the route, without the need to build a tram loop.

4.3 Third Step: Use on Shortened Routes

Passenger traffic in cities varies in many respects. Meeting the growing transportation needs requires integrated actions involving not only infrastructure development, but primarily changes to the traveling behavior of residents in the appropriate modal split, which is a significant share of journeys by public transport [21].

The adjustment of communication lines to the size, directions, and time of passenger flows is particularly important in urban transport. A tram system using bi-directional trams gives the possibility of a more flexible adjustment of the course and length of communication lines. Passenger traffic, increased during the rush hour, especially in the city center, can be handled by bi-directional vehicles for additional, shortened or changed lines within 24 h. The present tram system in Szczecin does not provide such a possibility. Each time the vehicle has to finish the line in a circular motion on the tram loop, it causes significant infrastructure occupation, extended tram travel time, and limited possibilities to increase the frequency of running vehicles on a given line. As a result, increased transport needs of residents are met by bus vehicles. The bi-directional tram could therefore be a tool for maintaining a balanced modal split within public transport in the city.

Figure 3 shows an example of the use of a switch-back for the implementation of a shortened route by the bi-directional tram in Szczecin.

Ongoing or planned investments in the tram system in Szczecin include several aspects: maintenance, repair and modernization of existing infrastructure, construction of new infrastructure sections, ongoing repairs and maintenance of tram rolling stock, purchase of a new tramway fleet, ongoing repairs and maintenance of the traction network, and construction of new elements. The rolling stock investments planned by TS and infrastructural projects designed by the municipality would allow for the introduction of the bi-directional tram in Szczecin in the near future.

Moreover, the implementation of the investment, including the introduction of bi-directional trams into the system, may be associated with measurable economic benefits. Rolling stock manufacturers point to differences in the price of vehicles. In their opinion, these differences can range from 8 to 15% to the detriment of the bi-directional rolling stock. However, the analysis of finalized orders in the Polish market indicated that these differences were often lower, reaching even a mere 4.5% (for the city of Poznań). It should be noted that the given values cannot be accepted as a fixed conversion rate, any purchase cost of a tram rolling stock depends on the requirements of the ordering party and the number of ordered vehicles. The indicated difference in the purchase price can be compensated by: limiting the demand for areas for surface-consuming tram loops, the possibility of tramway construction

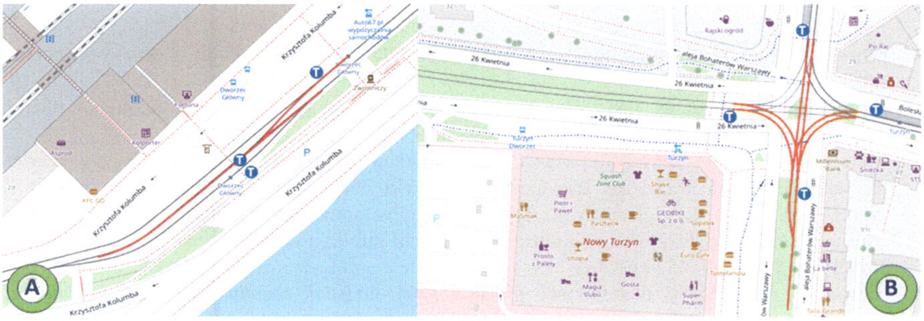

Fig. 3 The proposed use of switch-back for handling a shortened route by a bi-directional tram in Szczecin, (**a**) Kolumba St. (**b**) Boh. Warszawy/Krzywoustego intersection. (Source: Own work based on www.openstreetmap.org)

stages with using particular fragments, more comprehensive use of tram rolling stock, the promotion of environmentally friendly tram transport, and effective meeting of transport needs of residents.

5 Summary

The tram system is a part of the public transport system in cities. Considering the characteristics of vehicles and tram infrastructure, its operation should be consistent not only with the objectives of the transport policy, but also the spatial policy of a given area. The use of bi-directional vehicles in an urban tram system offers measurable benefits in terms of handling the passenger traffic, competitiveness of public transport, and spatial development of the city. It also offers the possibility of implementing the assumptions of sustainable development of public transport in the city.

The gradual implementation of the bi-directional tram into the Szczecin transport system suggested by the authors offers, among others, the following benefits:

- Increasing the functionality of the tram system
- Facilitating the traffic during tram line renovations
- Limiting the use of buses as emergency transport services
- Adjusting the tram network to the spatial development of the city
- Limiting the need to construct tram loops

All the aforementioned benefits also offer efficient passenger services and fulfil their transport needs.

The inclusion of bi-directional vehicles into the Szczecin tram system proposed by the authors is a solution which should be considered by other cities with tram networks, which experience difficulties with spatial urban expansion.

References

1. https://www.un.org/development/desa/en/news/population/2018-revision-of-world-urbanization-prospects.html
2. Putra, K.E., Sitanggang, J.M.: The effect of public transport services on quality of life in Medan City. Proc. Soc. Behav. Sci. **234**, 383–389 (2016)
3. Pietrzak, O.: System transportu pasażerskiego województwa zachodniopomorskiego w ocenie jego użytkowników. Problemy Transportu i Logistyki. **35**(3), 131–141 (2016)
4. Ngoc, A.M., Hung, K.V., Tuan, V.A.: Towards the development of quality standards for public transport service in developing countries: analysis of public transport users' behavior. Transp. Res. Proc. **25**, 4560–4579 (2017)
5. Sierpiński, G.: Miary dostępności transportowej miast i regionów. Zeszyty Naukowe Politechniki Śląskiej. Transport. **66**(1825), (2010)
6. Schmöcker, J.-D., Bell, M.G.H., Lam, W.H.K.: Importance of public transport. J. Adv. Transp. **38**(1), 1–4 (2003)

7. Fang, Y., Zeng, A.Z.: Long-term collaboration mechanism for disruption recovery service in public tram systems. Proc. Comput. Sci. **60**, 1337–1346 (2015)
8. Illustrated Glossary for Transport Statistics, 4th ed. Eurostat, European Commission. https://ec.europa.eu/eurostat/ramon/coded_files/transport_glossary_4_ed_PL.pdf (2009)
9. Regulation (EU) 2018/643 of the European Parliament and of the Council of 18 April 2018 on rail transport statistics. https://eur-lex.europa.eu/eli/reg/2018/643/oj
10. Rozporządzenie Ministra Infrastruktury z dnia 2 marca 2011 r. w sprawie warunków technicznych tramwajów i trolejbusów oraz zakresu ich niezbędnego wyposażenia, Dz. U. 2011, Nr 65, poz. 344
11. Zhou, Y., Qian, C., Xiao, H., Xin, J., Wei, Z., Feng, Q.: Coupling research on land use and travel behaviors along the tram based on accessibility measurement—taking Nanjing Chilin tram line 1 as an example. Sustainability. **11**, 2034 (2019)
12. Chen, C.L.: Tram development and urban transport integration in Chinese cities: a case study of Suzhou. Econ. Transp. **15**, 16–31 (2018)
13. Alpkokin, P., Kiremitci, S.T., Black, J.A., Cetinavci, S.: LRT and street tram policies and implementation in Turkish cities. J. Transp. Geogr. **54**, 476–487 (2016)
14. Sikorski, P., Wińska-Krysiak, M., Chormański, J., Krauze, K., Kubacka, K., Sikorska, D.: Low-maintenance green tram tracks as a socially acceptable solution to greening a city. Urban For. Urban Green. **35**, 148–164 (2018)
15. Dudkin, E.P., Smirnov, K.A., Doronicheva, S.A.: Justification and expansion of spheres of application of the tram. Proc. Eng. **189**, 848–853 (2017)
16. Bank Danych Lokalnych. https://bdl.stat.gov.pl/BDL/start
17. Pietrzak, K.: Analysis of the possibilities of using "Light Freight Railway" for the freight transport implementation in agglomeration areas (example of West Pomerania Province). Transp. Res. Proc. **16**, 464–472 (2016)
18. Plan zrównoważonego rozwoju publicznego transportu zbiorowego dla miasta Szczecin (2014–2025)
19. Zarząd Dróg i Transportu Miejskiego w Szczecinie. https://www.zditm.szczecin.pl/pl
20. Internal documents of Tramwaje Szczecińskie Sp. z o.o.
21. Bryniarska, Z., Zakowska, L.: Multi-criteria evaluation of public transport interchanges. Transp. Res. Proc. **24**, 25–32 (2017)

Customer Service Effectiveness in Shared Mobility Systems Using Artificial Intelligence Algorithms

Justyna Pawłowska

Abstract Sharing economy has been developing dynamically for a few years. Shared mobility systems are one of the most interesting dynamically developing concepts of the sharing economy. All around the world alternative transport modes have become an object of interest while being promoted as innovative forms of urban mobility. Shared mobility systems are focused on reducing the number of transport vehicles by using each of them more efficiently. In order for these systems to develop, an efficient customer service is needed. Nowadays, artificial intelligence is increasingly reshaping services by performing various tasks in an innovative way. Artificial intelligence algorithms for customer service are pushing the envelope of innovation and revolutionizing the way customers are assisted. Currently, AI-assisted bots can efficiently handle first-level queries and assist in reducing operational costs. Currently, artificial intelligence is used in customer support service in industries such as: finance, retail, travel, food and transport. The aim of the chapter is to create an algorithm dedicated to the customer service in shared mobility system which includes sample dialogs, flow diagram and prompt lists adapted to shared mobility systems.

Keywords Transport · Shared mobility · Artificial intelligence · Customer service

1 Introduction

Dynamic social, environmental and economic processes lead to an increase of mobility with the use personal cars. This leads to increased congestion, degradation of natural environment, a decrease in safety and thus the quality of life of citizens. Modern metropolises face a number of challenges. Most members of the society choose the personal cars to satisfy their mobility needs because they perceive it to be convenient, direct, safe and quick. At the same time, the political decisions indi-

---

J. Pawłowska (✉)
Faculty of Economics, University of Gdansk, Gdansk, Poland

M. Suchanek (ed.), *Transport Development Challenges in the 21st Century*, Springer Proceedings in Business and Economics,
https://doi.org/10.1007/978-3-030-50010-8_11

cate a determination to reduce the share of personal cars in everyday commute. However, in order for this goal to be achieved, there has to be an attractive alternative.

The services of shared mobility seem to be a possible answer to the transport demand of consumers. They are a part of a wider concept of sharing economy, a new model of socioeconomic relations. This model is based on the concept of resigning from ownership of goods in place of their shared use. The shared mobility services, while being different from each other, are commonly characterized by the desire to reduce the total number of vehicles being used. A number of different services have occurred on the market including: one car being used by multiple people at the same time (carpooling), rental services of generally available cars, motorbikes or scooters (carsharing, scootersharing) and systems of public urban bikes (bikesharing). Global as well as Polish experience indicates that these services are becoming more and more popular. Even though, different aspects of share mobility have been observable for many decades, they have developed radically due to the use of advanced information technologies.

An optimal level of customer service is needed for the strategic management of shared mobility companies due to the innovativeness, dynamic growth and high competitiveness at the market. Hence, more and more companies decide to use AI solutions, including conversational bots. The bots are a new type of customer interface which allow to use service and communicate with different solutions via a chosen communications service. There are: chatbots, conversational entities, conversational interfaces, etc. Once created, they are controlled by software and not by a human being. Bots are a novel type of communications tool with the customer and their existence affects the growth directions of modern programming the same way as mobile technologies and Internet did. This technical revolution led to the existence of services such as Uber, Traficar, AirBnB or Lime. The bots are used as a frontline of customer service and increase its efficiency. Depending on the level of algorithm, the bot can answer questions such as: "How to register and create an account?" or "In what areas can I use the shared cars/bikes?". Frequently asked questions (FAQs) are a natural direction of how the systems should be first created so that they are an automatic support of customer service personnel.

Artificial Intelligence provides the programmers with the tools necessary to create bots. Natural Language Understanding and conversation management tools facilitate creating complex text chats. Bots are more efficient, quicker and cheaper than a human being in recurring tasks [1]. Advanced technologies play a key role in the further development of shared mobility systems by changing the approach to customer service [2].

The chapter presents the role of technology in the support of customer service in shared mobility. The theoretical part has been based on current literature, including reports and statistics on share mobility, sustainable mobility and the development of cities and the use of artificial intelligence in customer service. The second part of the chapter presents a project of an algorithm to be used in a bot for shared mobility.

2 The Idea of Shared Mobility Services

Shared mobility is an innovative transport service, which allows its users to use chosen vehicles for a short term without the need to purchase them. Shared mobility includes the rental of cars, bikes, scooters and also a shared use of one vehicle (carpooling). The services of shared mobility have increased the transport accessibility of citizens while also decreasing the need to own a personal car. Most of the operators of share mobility services choose to integrate advanced technologies in their services so as to allow the customers to reserve the vehicles, track them and use the systems as a whole.

First bikesharing systems were implemented in Amsterdam in 1965. The goal was to provide the citizens with an access to alternative vehicles and to protest against the increased motorization within the cities which was troublesome both spatially and environmentally. 20,000 were supposed to be purchased but the project collapsed due to insufficient funding. Since then, the bikesharing systems have evolved significantly, including a rental charge and in the third generation of system smart locks—docking stations where the bikes can be left. There has been a significant technological development in terms of paying for the services and identifying the users. The control of users allowed to enlarge the areas in which the system works. Currently, telematic solutions are necessary to manage and control the number of bikes at different stations. Nowadays, fourth generation of bikes are introduced—so-called intelligent bikes—which are characterized by a system of identifying the users, renting, blocking, and returning the bikes. The bike itself performs the functions of the docking station because it has an electronic lock and a multiaspect tracking system which collects the data on a continual basis (including the possibility of GPS tracking). Renting a bike is now usually done with the use of a smartphone rather than the docking station itself which was common in the third generation of bikes [3].

Bikesharing systems exist in over 800 cities, in 50 countries all across the world. Almost a million bikes are docked in over 38 thousand docking stations [4]. The first table presents the largest bikesharing operators in Europe (Table 1).

Carsharing is a system of car rental per minutes. The first carsharing system was created in The Netherlands. In 1968 a one and only one electric carsharing vehicle was presented followed by a couple dozens of vehicles produced in the following years. The Dutch based their system on stations because they had a relatively small fleet of vehicles on hand. The system didn't work as planned because too many cars were in one place while other stations, at the outskirts of the system area were empty. The charging of the vehicles took too long and the driving distance per charge was too short. The system didn't work but it started a new era in the share mobility evolution.

Modern carsharing systems are fairly similar to traditional car renting, but they are different in a number of aspects [5, 6]:

- Cars are available 24/7—they can be reserved, shared and returned.

Table 1 Largest bikesharing companies in Europe in 2017

City	Operator	Name	Year of launch	No. of stations	No. of bikes
Paris	JCDecaux	Velib'	2007	1746	23,845
Brussels	JCDecaux	Villo!	2009	360	5000
Lyon	JCDecaux	Velo'v	2005	348	4000
Valencia	JCDecaux	Valenbisi	2010	275	2750
Sevilla	JCDecaux	Sevici	2007	260	2500
Toulouse	JCDecaux	VeloToulouse	2007	253	2400
Barcelona	C.Channel	Bicing	2007	424	6000
Milano	C.Channel	BikeMi	2008	257	4650
London	Bixi/Alta	Santander	2010	687	9200
Montpellier	Smoove	Velomagg'	2007	52	2414
Warsaw	Nextbike	Veturilo	2012	355	5100
Munich	Nextbike	Nextbike	2011	30	3000

- The system is characterized by self-service of reservation, sharing and returning.
- Rental time is charged per minute or hour depending on the operator.
- Advance information technologies are used.
- The area of the use of cars is preset (cars have to returned within the area).
- Insurance and petrol costs are included in the price of the service.

In Poland, the idea of carsharing is fairly new. It was first introduced in Warsaw in 2016 where it has grown dynamically. Recently, a number of new technologies were developed, which are supposed to ease the management of a carsharing system, improve the customer service and decrease the costs of implementation of further systems.

3 Artificial Intelligence in Automated Customer Service

An automatization of customer service increases the response quality to a level not available to a human being. AI-based systems can be used as prevention by monitoring the activity of customers in applications based on different danger indicators. They identify the customers which have problems while collecting data on the type and scale of the problems. Artificial intelligence is based on the process of self-learning of machines which thus present life-like, intelligent behaviour. The system can react in real time based on a support of virtual agents and databases of FAQs. This problem can be predicted and solved before it has even occurred thus increasing the satisfaction of the customer and decreasing the number of complaints and resignations [7].

Furthermore, employing and training employees is a time and cost consuming process. An increase of demand for the services leads to an increase of employment and thus, naturally, the employment costs. In the long perspective, the automatization platforms decrease the costs radically because the process of machine learning is self-preservative and require diminishing costs. Automated customer service works 24/7 and is not constraint by holidays and labour contracts. Problems are solved as they occur, on the go, without the need for the customer to wait. Such a system positively affects the satisfaction of the customers and the reception of services solved. The brand itself becomes perceived as more reliable and engaged in the customer's problems [8].

Customer service based on the AU algorithms provides a customer service level which is not available for a human being. Conversational bots are not based or emotional and their only goal is to provide an answer which is as good as possible based on the algorithms of service. The mode of communication between companies and their customers changes significantly. The Generation Y (Millenials) even avoids interaction with another human being and chooses self-service options [9]. In case of new generations, the system of customer service has to be as approachable as possible, otherwise the company might lose its customers.

New technologies allow the customer service to be dynamically developed. An introduction of a new product might be followed with an FAQ being published which allows troubleshooting for the most common problems. An assistance of AI in customer service allows a company to facilitate entering a new market reducing the training, financial, language and social barriers. The automatization of customer service thus allows a more dynamic development of a company.

The AI systems which are now used are characterized by a number of features which are crucial for their implementation at a shared mobility market which requires quick responses. The features include:

- Natural language understanding—the speech is transformed by an algorithm which allows to understand the intention of the customer and selects the most important key words from the speech. This process involves machine learning based on an analysis of dialogue (Fig. 1).
- Management of discussion—the dialogue with the customer is streamlined—people often cover different topics during discussion, often the goal of the discussion itself changes—artificial intelligence allows to select the right context and manage complex discussions with conversational bots.

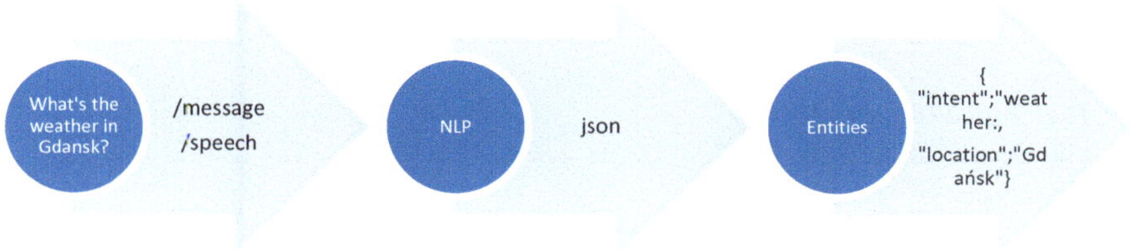

Fig. 1 Transcription of speech into intention

- Image recognition—the ability to recognize text, objects and even emotions on images based on training data—this function can be used when a customer sends over e.g. a photo of a receipt.
- Prediction—a correct reaction of a bot at a given moment of conversation—AI usually can predict the outcome of the conversation given a sufficient amount of learning data.
- Analysis of sentiment—the ability to recognize the emotion of the customer as well as the confidence level in the virtual agent.

The artificial intelligence provides the companies with the tools necessary to create conversational bots. The understanding of natural language and the discussion management tools are crucial to manage the conversation between the bots and customers.

4 Automatization of Damage Report in a Carsharing Company

The graph (Fig. 2) presents a structure of a process of damage report system in carsharing company (Traficar). Four parts of the dialogue can be distinguished:

- Dialogue steps (moves).
- Intentions (intet).
- Variables (slots).
- Actions.

The dialogue steps are moments of conversation in which the bot asks for the intentions of fills in needed information. They are presented as ovals within the graph. These include:

- Greeting and asking about the type of the problem.
- Asking for the PESEL number.
- Informing about an incorrect PESEL number and a request for repeating.
- Informing about a lack of possibility to verify the identity.
- Asking for the place of the accident.
- Asking for a short description of the accident.

The intentions are different requests, commands, questions, and answers of the users based on the questions of the Assistant. The intentions are indicated as the vectors within the graph

- The choice of "accident".
- The choice of "registration".
- The choice of "reservation".
- The choice of "payments and receipts".
- The provision of PESEL number.

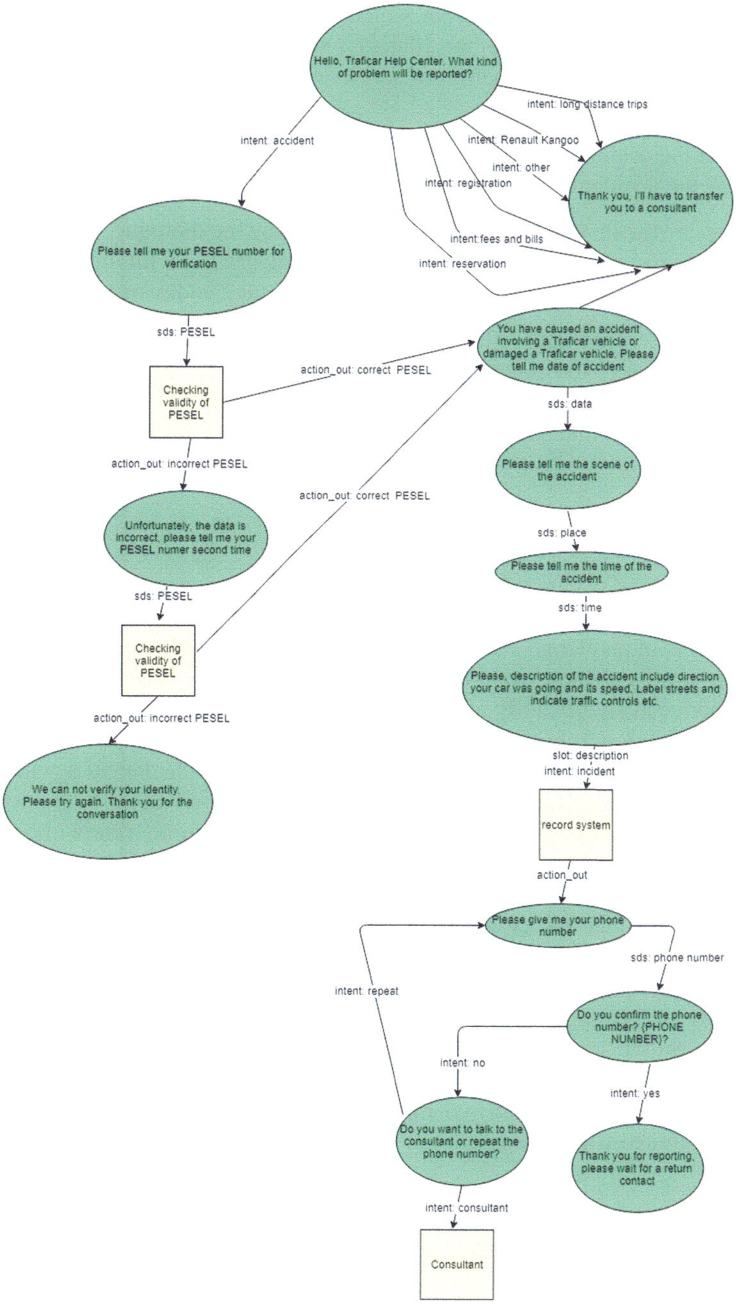

Fig. 2 Dialogue graph—an automatization of a part of damage report system

The variables (slots) are the data which can be collected from the intentions. Some of them are generic (numbers, dates). Others are variables which are specific for a given dialogue, e.g. the names of services or products:

- PESEL.
- Place of the accident.
- Date of the accident.

Actions are the operations on data collected during the dialogue. They are implemented for a given business process and require integrating with the system. In the analysed example actions are indicated by rectangular nodes in the graph. These include:

- Switching over to the consultant.
- Validation of the PESEL number.
- Saving the date, time and place of the accident in the system.

5 Conclusions

The development of urbanization presents the people responsible for urban mobility with a number of challenges. Shared mobility and intermodal transport solutions are becoming more and more significant for the cities. There are the global economic trends which can significantly alter the urban mobility, because right now the individual motorization ratio is growing and the accessibility of parking spaces is decreasing. The operators of shared mobility systems provide the cities with a solution which can complement public transport. The shared mobility systems are becoming popular among people who can abstain from owning a car without losing the comfort of driving one. However, in order for these systems to work in an efficient way, new information technologies have to be implemented. These applications allow to change the customer service approach and automatize some of the labour currently carried out by employees. Automatization is the future of customer service, especially in share mobility services, which, due to their specifics, require for example 24/7 control of the processes. The projects of the bots are rapidly evolving and innovative companies use them because they want to increase their productivity and create a new culture of customer service. Therefore conversational bots can and should be used in shared mobility companies to carry out the customer service in a more efficient way.

References

1. Shevat, A.: Designing Bots. O'Reilly Media, Sebastopol (2017)
2. Shaheen, S.A.: Applying Integrated Its Technologies to Carsharing System Management: A Carlink Case Study. Institute of Transportation Studies, UC Davis, Davis

3. Shaheen, S.A., Cohen, A.P., Martin, E.W.: Public bikesharing in North America: early operator understanding and emerging trends. Transp Res Rec J Transp Res. **2387**, 83 (2013)
4. The bike-sharing world map—Google My Maps. https://www.google.com/maps/d/u/0/viewer?ie=UTF8&hl=en&om=1&msa=0&ll=-3.81666561775622e-14%2C-42.890625&spn=143.80149%2C154.6875&z=1&source=embed&mid=1UxYw9YrwT_R3SGsktJU3D-2GpMU. Accessed 12 May 2019
5. Brychcy, M., Przybyłowski, A.: Funkcjonowanie i rola carsharingu w równoważeniu mobilności na przykładzie Trójmiasta. Studia i Prace Kolegium Zarządzania i Finansów. **169**, 43–56 (2018)
6. Shaheen, S., Cohen, A., Bayen, A.: The Benefits of Carpooling. Transportation Research Board, Washington DC (2018)
7. Rubin, V.L., Chen, Y., Thorimbert, L.M.: Artificially intelligent conversational agents in libraries. Libr Hi Tech. **28**(4), 496–522 (2010)
8. Zakos, J., Capper, L.: CLIVE—an artificially intelligent chat robot for conversational language practice. In: Darzentas, J., et al. (eds.) Artificial Intelligence: Theories, Models And Applications, pp. 437–442. Springer, Berlin (2008)
9. Suchanek, M., Szmelter-Jarosz, A.: Environmental aspects of generation Y's sustainable mobility. Sustainability. **11**(11), 3204 (2019)

Poland as an Example of Using Flexible Transportation Systems as a Tool for Meeting the Needs of Mobility

Tomasz Kwarciński

Abstract The article presents theoretical and pragmatic aspects related to the organisation and operation of the flexible transportation systems. Such systems are seen as promising from the point of view of ensuring mobility to the inhabitants, especially those living in the areas with a low demand for transport services. After 1990, inhabitants in multiple regions across Poland were deprived of their access to public transport. The main reason for which such services were no longer provided was the decrease in demand for public transport. Flexible systems may become a valuable supplementation of the public transport, as well as an alternative for individual car transportation, in the pragmatic level, the functioning of the flexible transportation in the city of Szczecin. This city is one of the two in which the transport on demand has been introduced. The statistical data analysis makes it possible to see the increased use of this form of transportation by the inhabitants.

Keywords Flexible transport systems · Public transport · Mobility

1 Introduction

Within the market, the transportation requirements of the inhabitants can be met by providing regular and irregular transportation services. The characteristic feature of the regular transportation service is a schedule that is available to the public, which does not apply to the irregular transportation service. The irregular transportation service is applied only after the requirement for it is submitted (e.g. via phone call) by the individuals that will use such a service. This manner of fulfilling the transportation needs of the inhabitants is referred to as a flexible transportation. The avail-

T. Kwarciński (✉)
Department of Transport, Faculty of Management and Economics of Services, Szczecin University, Szczecin, Poland
e-mail: tomasz.kwarcinski@wzieu.pl

M. Suchanek (ed.), *Transport Development Challenges in the 21st Century*, Springer Proceedings in Business and Economics,
https://doi.org/10.1007/978-3-030-50010-8_12

able literature also refers to such a solution as a demand-driven transportation services. In practice, the term "transportation on demand" is also used.

The flexible transportation fulfils two primary objectives. The first has a social nature and it is related to providing the inhabitants with access to public transport. The second objective has an economic nature and is related to the rationality of the using: means of transport, limited resources and minimising the costs of the transportation services provided.

The purpose of this article is to present the theoretical and practical aspects regarding the function of the flexible transportation system. This issue has been rarely described by any literature related to the subject. On the theoretical level, attention was drawn to the flexibility of the public transport system, highlighting its limitations and benefits. However, within the pragmatic aspect presents the premises of the development regarding this form of public transportation as well as the selected aspects of organisation and functioning of the flexible transportation as it is implemented in Szczecin. The article presents statistical data obtained from the organiser of the flexible transport in Szczecin—the Public Transport Authority.

2 The Essence of the Flexible Transportation System

Flexible transport systems constitute a new type of passenger transport services, the so-called demand-driven services. They are a form of public service implemented in response to individually/collectively reported transport needs (so-called services on request, or demand). Services of this type are provided in the times of days characterised by low demand for public mass transport services. They can be used as: main offer in medium-sized towns, with extension to neighbouring towns and rural areas, transport to a regular transport line, basic service network in rural areas, complementary services (e.g. on non-working days), dedicated or special services [1, 2].

Irregular transport in the literature of the subject is considered promising for areas with low demand for transport services [3]. They constitute one of the solutions applied to the problem of limited (or lack of) access to passenger transport in rural areas [4]. In addition, like the regular public transport, they are an important alternative to individual car transport [5].

Within the framework of the flexible transportation systems, we can distinguish between the following transportation services being provided [5, 6]:

- on communication lines,
- within a specific area,
- by taxi services (companies),
- for people with limited mobility (special transportation),
- by means of transportation provided to the inhabitants.

A characteristic feature of the flexible transportation is the irregularity, which regards the time and the area of the transport services. These two parameters allow to distinguishing between the two basic types of transport on demand: linear and area.

The linear system is included in the schedule, but it is used upon request. Its characteristic feature is the utilisation of a predetermined communication line and stopping only at stops specified by the passengers. This system may also have a mixed nature, i.e. some of the routes on a given line are served regularly (most often school and employee transport), and some routes are carried out only on demand, most often at lunchtime or in the afternoon. The time of utilising the flexible system is usually limited to working days. The flexibility may also apply to the period in which the transport services are provided. For example, during holiday time, holiday breaks, a portion of regular routes is altered into irregular ones.

The time in which the transport services are provided is determined by the passengers themselves, which means that the role of a potential passenger in organising the transportation is increased. This influences the level of adjustment to the needs of the inhabitants, which are significantly smaller in rural areas. The flexible transport services provide the opportunity for a better adjustment between the time in which such services are required and the provision of the said services. The growing demand for transport may prompt the transport organiser to implement a transport on demand within the timetable, provided that the operator will perform the service upon receiving prior notification. This type of adjustment lies within the linear transportation system on demand.

The irregularity of flexible transportation, unscheduled, may also apply to the planned routes. This type of organisation is related to the flexible transport area system, which is characterised by the lack of predetermined communication routes. The route of the vehicle is determined by a special program (algorithm) or individually by the dispatcher utilising the flexible transportation system in contact with the driver. In this system, it is important to establish a transport point, to which the passengers from a given area are transported. Passengers travelling a longer distance can use regular transport with the ticket they had purchased which allows them to meet their demand. In addition, in this system, it is necessary (if the organiser decides so) to introduce additional fees which would be added to the total price for the transport service.

The flexible transportation system allows the performance of multiple functions regarding a social, economic and environmental nature. This form of transport is used to improve the level of accessibility to public transport services which is significant in areas with limited access or with no access to public transport. In social terms, it allows limiting the feeling of social exclusion.

Public transport plays an important role related to the Public Service Obligation (PSO) [7]. In social terms, this form of transport also contributes to limiting the phenomenon of social exclusion [8]. In economic terms, public transport enables residents to move around, taking into account the spatial distribution of production and settlement centres [9]. By enabling access to public transport for people who, due to their age, predispositions and economic status, are unable to use other forms of transport (e.g. using individual car transport), give opportunities to satisfy all obligatory and optional needs (work, school, shopping, health, culture, social purposes) [10].

From the economic point of view, the issues related to the profitability of the flexible transportation system are important. In general, it can be assumed that profit is not the goal of providing flexible transportation services. The transportation services are treated as a public good that requires subsidising. However, it should be noted that the cost of providing transport services in a flexible manner is lower than providing such transportation services on a regular basis. Financial support may be used for the purchase of various means of transport and the cost of their maintenance by public authorities. The important and, at the same time, expensive part of the management system is the one (the so-called recording organisation) which by using the appropriate software, the management of the means of transport, establishing the routes, collects information regarding the demand for transit, etc. Another type of cost, which, to a certain extent, applies to the regular transportation, is the so-called readiness of the means of transport to provide a service and that includes driver's wages and the maintenance of the vehicles that are in reserves.

The organisation of the irregular transportation takes into account the principle of rationality in the manner in which such services are provided. It is connected with the minimisation of costs required to satisfy the demands for transportation. The cost of providing irregular transportation services is lower than the maintenance of the regular transportation, especially in areas characterised by low demand for transportation services.

The development of the flexible transportation systems requires a support from public authorities in terms related to the organisation of it. It is important to incorporate flexible system solutions into the general public transport system through integration of technical aspects, the time and the tariff. It gives a sense of stability of the transport system, creating a higher added value for passengers.

An important feature which distinguishes the flexible transportation systems is the use of ICT-based solutions (primarily telephone and the Internet) as well as mobile applications. They combine customers and transport operators while, at the same time, fulfilling many functions, e.g. providing information, requesting a means of transport, payments, etc. Maintaining traditional phone calls is especially important for elderly customers who often do not have access to the Internet or do not use it and do not use mobile applications which are available on smartphones.

In some types of flexible systems, there is the possibility to use the means of transport in a better way by directly stating the destination. In this manner, the direction of transit is determined. It can be, for example, a shopping centre, as well as reaching a destination related to receiving healthcare services (e.g. reaching any hospital).

3 The Premises for the Development of Flexible Transportation Services

The provision of transportation services on a regular basis is typical for areas with high demand for transportation services, which affects the frequency in which the said services are provided. For the transport user, the frequency in which the transportation services are provided is the deciding factor when choosing the method of travelling. The higher frequency in which the services are provided also increases the attractiveness of public transport.

The situation is different in the case of areas which are characterised by small and dispersed transportation demand as it generates low demand for transport services. In such areas, transportation services are provided rarely—several times a day or less. In many cases, the consequence of the low (and decreasing) demand for transportation services is that the entities responsible for organising the transportation services decide to limit the frequency and the offer of such services, and on the other hand, the lack of the said offer reduces the needs or raises the need to search for a substitute to such services in order to travel.[1]

Restricted access to transportation services may regard both the time and the location. As a result of that, the attractiveness of public transport decreases, especially in comparison to a passenger car, which reduces the demand for public transportation services. Apart from an individual car transport, the demand for transportation can also be met by utilising various forms of travel sharing, e.g. *car sharing* or *carpooling*.

In Poland, limiting the supply of transport services is particularly visible in the nonurban areas and it is correlated with the dynamic development of individual car transport in such areas. Passenger cars in extra-urban areas are usually old, in a worn technical condition, and more detrimental to the environment than newer cars. However, they allow their users to meet their transportation demands. Therefore, they play a positive and important role for the inhabitants of these areas, although from a social and environmental point of view, they generate detrimental results. It is due to the manner in which they are used and recycled [11].

In Poland, between the 1980s and 1990s, the number of transported passengers exceeded 2 billion per year with an automotive index of 100 vehicles per 1000 inhabitants. It should be emphasised that the beginning of the 1990s was a time of big changes within the passenger transport market, in particular caused by its liberalisation. The effect of the changes was the introduction of a large number of small enterprises onto the market, which due to the small number of people employed were not subject to the statistical reporting obligation. Changes in the length and number of communication lines, transport volumes, automotive index and participation of individuals with a passenger car are shown in the Table 1.

The economic changes in Poland, which took place after 1990, had an impact on the decline in regular bus transport (as well as rail transport).[2] The largest decline was recorded in the first years during which the said changes took place in Poland. In the years 1990–1995, the decline in passenger transport by regular bus transport was equal to 46%. In the period 1990–2017, the decline in the number of passengers

[1]This is the case for public transport services provided in nonurban areas. In Poland, they are implemented on a commercial basis, which means no subsidies from the state or local government units. Transport operators are also the organisers of public transport in this case.

[2]In rail transport, the number of transported passengers amounted to 789 thousand. (1990) and 303 thousand (2017) [12, 13].

Table 1 Activity of the regular bus transportation in Poland between 2003 and 2017 within the scope of changes in the individual automotive index

Year	Length of regular bus lines (in thousands of km)[a]	Number of regular bus lines (in thousands)	The volume of transport by the regular means of transport (in millions of pas.)[b]	Total individual automotive index (number of passenger cars per 1000 inhabitants)	Participation of individuals with passenger cars in urban areas (in%)	Participation of individuals with passenger cars in nonurban areas (in%)
2003	1236.4	26.7	688.6	294	44.0	51.0
2005	1284.1	25.3	671.1	323	44.7	–
2007	1234.5	24.1	609.4	383	52.5	–
2009	1110.1	22.2	516.7	432	55.8	67.2
2011	947.6	18.5	446.4	468	56.6	67.6
2013	838.7	16.3	385.3	504	56.5	70.9
2015	754.0	14.3	347.7	539	58.2	71.1
2017	627.5	13.0	312.1	586	60.1	75.3
2017: 2003	−49%	−52%	−55%	+99%		

Source: based on our study of the data provided by: Transport wyniki działalności 2003, 2005, 2007, 2009, 2011, 2013, 2015, 2017, GUS, Warszawa, 2004, 2006, 2008, 2010, 2012, 2014, 2016, [14] and Sytuacja gospodarstw domowych w 2003, 2005, 2007, 2009, 2011, 2013, 2015, 2017 r. w świetle wyników badania budżetów gospodarstw domowych, GUS, Warszawa, 2003, 2005, 2007, 2009, 2011, 2013, 2015, 2017

[a]Of enterprises employing more than nine persons, providing regular national and international traffic; excluding urban transport services. Data concerning line of scheduled bus communication includes: suburban lines, regional lines, long-distance lines

[b]In enterprises employing more than nine persons. Data on road transport of passengers does not include transport of persons by buses: of urban public transport enterprises, enterprises carrying out tourist services and transport of persons by buses on own account

using bus transport in Poland was equal to 85%. The drop in transport had an impact on further limitations of the length and number of communication lines. According to the statistical data, in 2003–2017 the length and number of communication lines has decreased by half in Poland.

The decline in the regular bus transport in Poland occurred for various reasons. As previously mentioned, within the initial period, it was associated with the changes in the economy. The fall of large enterprises and the development of micro enterprises resulted in changes in the scale, structure and the route of vehicles. The decline in passenger transport was also due to the increasing number of passenger cars in Poland. Since 1990, there has been a large inflow of used cars to Poland, which, in some respects, was a response of the residents to the limited access to public transport.

The increase in the number of passenger cars had an impact on the growth of the automotive index and the participation of individuals with passenger cars. Currently, the automotive index in Poland has reached values that are present within the

Western European countries. However, the overall participation of individuals with a passenger car exceeded 60%. The dependence based on a sustained higher participation of individuals with passenger cars located in nonurban areas should be noted. In addition to the higher volume of passenger cars, the suburban areas in Poland are characterised as higher, in relation to urban areas, number of vehicles: scooters, motorcycles and bicycles. This may be due to the fact that residents of rural areas have to travel longer distances for travel purposes.

The continual advantage based on the participation of individuals with passenger cars in nonurban areas within Poland should not be combined with their higher economic status but rather with a practical approach to the ability to meet transportation needs. In view of the continuous reduction of the length and the number of regular communication lines in rural areas, a passenger car provides a better mobility options for these inhabitants.

The process of limiting the length and number of communication lines adversely affects the level of accessibility of public transport in nonurban areas. The development of individual motoring favours the attractiveness of public transport, which is reflected, among other things, in the decline in the number of passengers. This process can be described as a vicious circle, which in extreme cases leads to an integral disappearance of access to public transport and, consequently, creates a sense of social exclusion. In Poland, this phenomenon can be observed in nonurban areas. In the light of the research carried out by the author, the disappearance of the offer of transport services occurs throughout the country, although the majority of nonurban locations without access to public transport occur in eastern Poland. In this area about one-third of rural areas are deprived of access to public transport. In addition, the factors that limit the attractiveness of public transportation services within nonurban areas in Poland are: low frequency of services and lack of up-to-date timetables placed in the paper version at transport stops.

According to other studies [11] referring to the level of accessibility of public transport in non-urban areas in Poland, there is a relationship between the development of the communication network and the settlement structure. The regions of eastern and central Poland are characterised by lower availability of bus transport services. From the point of view of the settlement structure, suburban settlements of this area are smaller and more dispersed. In contrast, in southern and western Poland, the communication network is better developed. Suburban villages located in this area are larger and are characterised by a compact design.

The drop in demand for transport services was also due to the fact, that the state stopped subsidizing such services. To a certain extent, compensation for residents of nonurban areas in Poland was an extension of the scope of statutory concessions. However, limiting the availability by commercial carriers resulted in the lack of the possibility of using the rights granted to the residents.

The flexible transportation system could improve the situation in suburban areas within Poland, especially those with no access to public transportation services. This form of transport can be considered as important for the revitalisation of public transport in areas without access to it [4]. It is impossible to expect that a car will replace public transport. Not all people, for various reasons, may own and also use without the support of other people from this mode of transport (children, adolescents, the elderly, no entitlements).

4 The Organisation and Functioning of the Flexible Transportation System in Szczecin

Szczecin is the second city in Poland, which introduced flexible transport into the offer in April 2016. Its purpose was to provide the inhabitants with the regular means of transport that would operate within the peripheries and low density areas to their destinations.

The organiser of the transportation services on demand is the Transport Management Office in Szczecin. This entity, as the municipal budgetary establishments, is responsible for the organisation of public transport within the city and in the municipalities which, on the basis of inter-municipal agreements, requested the organisation of local transportation service in their area (Kołbaskowo, Dobra).

The role of the flexible transportation organiser is to choose the operator. The operator's experience in transport and the price of the service influence the choice. The selected operator receives from the organiser the payment for the provided transport services, which includes two components: the actual kilometres travelled and the readiness to provide services. The organiser also covers the costs related to the booking of courses for which he is responsible located at the board of the Dispatching Centre.

The role of the Centre can be determined by their functions, which are: management of the transportation needs of the inhabitants requesting transportation services on demand, planning and managing the system including the vehicles. Its function is also to maintain contact with flexible transport services providers.

Initially, the flexible transportation services were limited to one of the districts of Szczecin—Podjuchy. This location was chosen based on the features of that area, such as topographic conditions (hilly terrain), narrow streets and a relatively low population density (predominantly single-family houses). The transport volumes for this area are presented in Table 2.

Within 2.5 years of flexible transport operation in the Podjuchy area, nearly 95 thousand of the transport services have benefited from this form of transport services passengers. The relatively short period of functioning of flexible transport does not allow for the identification of significant dependencies (e.g. seasonality on service demand). In a detailed analysis of the functioning within the system, information on the structure of transport, transportation needs, etc. would be valuable. In general, the tendency associated with the growing number of people using this form of transport should be positively assessed.

From September 2018, the offer of the flexible transportation services was extended to the next district—Gumieńce. Two new flexible transport lines were launched in this area (98, 99). Since the launch of the 98 line, 986 (September

Table 2 Number of passengers using on-demand transport on the line 95

2016	Number of passengers	2017	Number of passengers	2018	Number of passengers
January	–	January	2429	January	2991
February	–	February	2082	February	3010
March	–	March	3026	March	3692
April	–	April	2591	April	3432
May	–	May	2966	May	3704
June	–	June	2799	June	5111
July	1136	July	2925	July	4085
August	1612	August	2968	August	3870
September	2257	September	4297	September	3799
October	2489	October	4639	October	4141
November	2512	November	5012	November	2589
December	2407	December	3023	December	3096
Total	*12,413*	*Total*	*38,757*	*Total*	*43,520*

Source: based on our study of the data provided by the Transport Management Office in Szczecin

2018), 1396 (October 2018), 1803 (November 2018) and 2619 (December 2018) passengers benefited from flexible transport services. On the other hand, on route 99: 1403 (September 2018), 1764 (October 2018), 1671 (November 2018) and 1584 (December 2018) passengers.[3] Taking into account the line in Podjuchy, jointly in Szczecin, in 2018, the flexible transportation services have transported over 57 thousand passengers.

In the initial period, the organisation of flexible transport in Szczecin was dynamic. The passenger determined the time of the journey. In addition, special software allowed the selection of the optimal route to/from the interchange point with regular transport. As a result of the growing demand for transport, the rules of its organisation have been modified. A timetable has been introduced, provided that the route will be established when such service is requested and the route of the vehicle has been planned. Taking into account the types of flexible transportation, it can be noticed that the growing demand for transport services forced Szczecin to switch from the area system to the linear system.

The communication line 95 (Podjuchy) launched in 2016 had initially 14 stops, later this number was increased to 16. It is serviced by 2 microbuses, and the system's operating time is limited to business days. Services are provided between 06:30 and 17:00. In turn, on lines 98 and 99 there are respectively 31 and 27 bus stops. The number of microbuses servicing them is the same as in the case of the line 95 after 02:00. Otherwise, the time of system operation is shaped. In the case of line 98, it is available on weekdays from 06:00 to 22:00, while line 99, in addition to providing services on working days at the same hours, also provides services on public holidays from 09:00 to 19:00.

[3] The data provided by the Transport Management Office in Szczecin.

5 In Summary

Flexible systems play an important role in suplementing regular public transport. They improve the accessibility of transportation services, which is important in areas characterised by a low demand in services. In social terms, they play an important role in preventing the emergence and deepening of the phenomenon of social exclusion. They have a positive impact on reducing the negative effects of transport on the natural environment. The development of this form of transport is an alternative to the development of individual car transport and the need to own a car.

The launch and development of the flexible transportation system in Szczecin should be assessed positively. The growing demand prompted the organiser of the flexible transportation system to establish new communication lines and include them in the timetable, provided that the service will be provided after its booking. For passengers, it is advantageous to be able to mix the tariff system already existing within the city, and new stops have improved the overall accessibility to public transport in the city.

References

1. Mężyk, A.: Elastyczne formy transportu publicznego w obsłudze komunikacyjnej regionu, w: Współczesne uwarunkowania rozwoju transportu w regionie, red. M. Michałowska, Studia Ekonomiczne. Zeszyty Naukowe Uniwersytetu Ekonomicznego w Katowicach, 143, Uniwersytetu Ekonomicznego, Katowicach. 266 (2013).
2. Withe, P.: Designing flexible transport services: guidelines for choosing the vehicle type. Transport. Plan. Technol. **36**(1), (2013)
3. Mageean, J., Nelson, J.D.: The evaluation of demand responsive transport services in Europe. J. Transp. Geograp. **11**, 255–270 (2003)
4. Mulley, C., Nelson, J.D.: Flexible transport services: a new market opportunity for public transport. Res. Transp. Econom. **25**, 39–45 (2009)
5. Velaga, N.R., Nelson, J.D., Wright, S.D., Farrington, J.H.: The potential role of flexible transport services in enhancing rural public transport provision. J. Public Transp. **1**(15), 111–131 (2012)
6. Kwariński, T.: Theoretical and pragmatic aspects of functioning of flexible transport systems. Eur. J. Service Manag. **20/1**(4), 159–164 (2018)
7. Szyszczak, E., Bekkedal, T., Davies, J., Andenæs, M.: Developments in services of general interest. Springer, Berlin (2011)
8. Grieco M., Urry J.: Mobilities: new perspectives on transport and society. Routledge Taylor and Francis Group, London and New York (2012)
9. Cebolada, A.: Mobility and labour market exclusion in the Barcelona Metropolitan Region. J. Transp. Geograp. **17**(3), 226–233 (2009)
10. Kenyon, S., Lyons, G., Rafferty, J.: Transport and social exclusion: investigating the possibility of promoting inclusion through virtual mobility. J. Transp. Geograp. **10**, 207–219 (2002)
11. Stanny, M., Rosner, A., Komorowski, Ł.: Monitoring rozwoju obszarów wiejskich. Etap III. Struktury społeczno-gospodarcze, ich przestrzenne zróżnicowanie i dynamika. EFRWP, IRWiR PAN, Warszawa (2018)
12. GUS: Rocznik statystyczny RP 2001, p. 407. GUS, Warszawa (2001)
13. GUS: Rocznik statystyczny RP 2018, p. 562. GUS, Warszawa (2018)
14. GUS: Transport wyniki działalności. GUS, Warszawa. (Transport—activity results). Statistics Poland, Warszawa) (2018)

Analysis of Trends in Development of Electromobility in Poland: Current Problems and Issues

Elżbieta Macioszek

Abstract The notion of electromobility spans an overall body of problems related to the use of electric vehicles (EV), covering both technical and service-related aspects of the EV, the charging technology and infrastructure, as well as social, economic and legal problems involved in the designing, manufacturing, purchasing and using of electric vehicles. The article addresses such matters as the share of electric vehicles in the automotive market as well as the problems of operation of electric vehicles both by private users and in public transport. Moreover, an analysis of trends in the development of electromobility in Poland, including current problems and issues, has been presented in this chapter. Some results of an analysis of accessibility of charging points and charging stations for electric vehicles currently used in the Conurbation of Upper Silesia and Dąbrowa Basin have also been discussed in the chapter.

Keywords Electromobility · Electric vehicles · Accessibility analysis

1 Introduction

The notion of electromobility spans an overall body of problems related to the use of electric vehicles (EV), covering both technical and service-related aspects of the EV, the charging technology and infrastructure, as well as social, economic and legal problems involved in the designing, manufacturing, purchasing and using of electric vehicles.

In 2017, the Polish Government adopted the Electromobility Development Plan [1] Aimed to create adequate conditions to enable development of electromobility and the industry related to this sector in Poland as well as stabilisation of the power

E. Macioszek (✉)
Faculty of Transport, Transport Systems and Traffic Engineering Department, Silesian University of Technology, Katowice, Poland
e-mail: elzbieta.macioszek@polsl.pl

M. Suchanek (ed.), *Transport Development Challenges in the 21st Century*, Springer Proceedings in Business and Economics,
https://doi.org/10.1007/978-3-030-50010-8_13

supply network. According to the plan, the electromobility development should proceed in three phases, depending on the degree of the market maturity and the required involvement of the state. One of the plan's assumptions was that the first phase would be of preparatory nature and last until the year 2018. This phase was planned to cover building conditions for the electromobility development in terms of the regulatory framework and establishing rules of public financing. In the second phase scheduled for the years 2019–2020, the infrastructure needed for charging of electric vehicles will be built in selected urban areas and incentives to buy electric vehicles will be intensified. The expectations are that results of the research activities performed in the area of electromobility, commenced already in the first phase, will be commercialised, and new business models aimed to popularise the EV will be implemented. In phase three, spanning the years 2020–2025, the electromobility market is assumed to have matured, which will allow for the public aid instruments to be gradually withdrawn.

The year 2018 saw the enactment of the act on electromobility and alternative fuels (Journal of Laws of 2018, item 317) [2]. The rationale behind the law was to regulate the market of alternative fuels in Poland, particularly with regard to electricity and natural gas. The act was intended to solve numerous problems, including the lack of sufficient infrastructure in agglomerations, densely populated areas and along trans-European transport corridors as a means to enable unconstrained flow of vehicles running on these fuels. According to its pre-assumptions, the act is expected to establish a regulatory system making it possible to build the public infrastructure required for charging of electric vehicles as well as for fuelling vehicles with CNG and LNG. All the related investments are supposed to contribute to the development of low-emission and zero-emission transport, which in turn is intended to reduce air pollution. The act provides for a system of incentives, including abolition of the excise tax on electric cars, increased depreciation charges for businesses, exemption from parking fees or admission to bus lanes for electrically driven vehicles. The document also proposes to introduce an obligatory share of electrical vehicles in the fleets managed by some of central administration bodies and selected local government authorities. The Parliament was proceeding very quickly on the act. Such haste was explained by representatives of the Ministry of Energy by the necessity to implement European Parliament Directive 2014/94/EU of 22 October 2014 on the deployment of alternative fuels infrastructure in the local legal framework.

Parallel to the preparations of the strategic electromobility documents performed over the recent years in Poland, numerous domestic research centres and scientific institutions have been studying the electromobility problem, producing a number of scientific publications elaborating upon electromobility in individual transport (including [3–13]) as well as in the sphere of public collective transport (including [14–19]).

This article addresses such matters as the share of electric vehicles in the automotive market as well as the problems of operation of electric vehicles both by private users and in public transport. Moreover, an analysis of trends in the development of electromobility in Poland, including current problems and issues, has been presented in this chapter. Some results of an analysis of accessibility of charging points and charging stations for electric vehicles currently used in the Conurbation of Upper Silesia and Dąbrowa Basin have also been discussed in the chapter.

2 Market Share of Electric Vehicles

An electric vehicle is a motor vehicle featuring a power transmission system containing at least one non-peripheral electric device functioning as an energy converter with an electrically charged energy storage unit which can be charged from an external source. Electric vehicles can be divided into:

- PHEV (plug-in hybrid electric vehicles)—automotive vehicles featuring a combined combustion and electric motor system, equipped with batteries that can be charged from external units.
- BEV (battery electric vehicles)—automotive vehicles that only use the energy accumulated in batteries for propulsion, while the batteries can be externally charged or replaced.
- EREV (extended-range electric vehicles)—electric vehicles with a built-in low-capacity engine unit. The latter can function as an external electric power generator when the vehicle's on-board batteries are discharged. Once it has switched on, the built-in engine functions as a typical power generator feeding the empty batteries.
- FCEV (fuel cell electric vehicles)—electric vehicles which, instead of the battery, use fuel cells alone or in combination with a battery or a supercapacitor to power the built-in electric motor. The fuels cells on board these vehicles generate electric energy to feed the motor typically by using airborne oxygen and compressed hydrogen. Most fuel cell vehicles are classified as zero-emission vehicles producing water and heat only. Compared to combustion engine vehicles, those that run on hydrogen accumulate pollutants in the hydrogen production unit, while the hydrogen itself is typically produced from reformed natural gas.

Electric vehicles are about to play a significant role in the transformation of both the transport and the power industry. They are currently considered as a prospective alternative for combustion engine vehicles. The potential benefits attributed to using them have partially triggered certain actions aimed at further development of electrification of car transport. The declaration signed at the Climate Summit in Paris in December 2015 stipulates that the share of the EV in the overall global stock of vehicles in operation should reach at least 20% by 2030 [20], while the forecast of the International Energy Agency (IEA), as provided in its annual report entitled "Global EV Outlook 2017" [21], is that the said share will rise to 30% by 2030. In the EU, the share of the PHEV in the overall volume of sale of new cars is assumed to come to 20% by 2030, while that of the BEV to 30%. Electric vehicles are ready-to-use solutions capable of helping the EU in its pursuit of the goals defined in the Paris Agreement of 2015. The automotive industry has already started investing in electric drives, and the number of customers who decide to buy such vehicles is growing on a regular basis. According to the data provided by the International Energy Agency [21], the number of electric vehicles (BEV, PHEV, FCEV) reached 3.1 million units globally in 2017, which corresponded to an increase of 54% compared to 2016.

In 2017, European customers purchased 216,566 EVs, which was 39% more than a year before (155,757 pcs.). Most EVs were sold in Germany (54,617 pcs.), the United Kingdom (47,298 pcs.) and France (36,835 pcs.). The number of plug-in EVs per one normal power charging position and one high power charging position respectively in selected countries in 2018 has been presented in Figs. 1 and 2. The

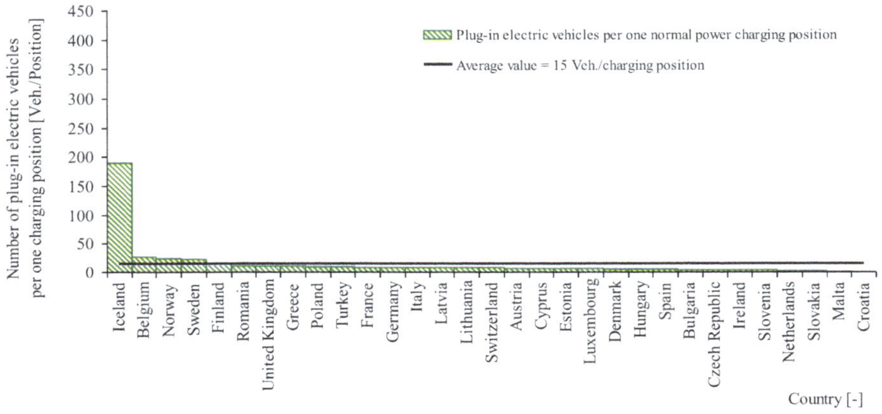

Fig. 1 The number of plug-in electric vehicles per one normal power charging position in selected countries in 2018 (Source: Own base on [22])

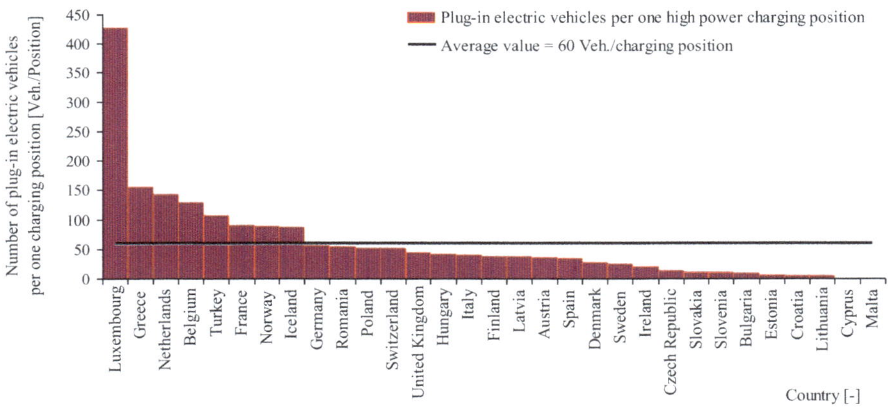

Fig. 2 The number of plug-in electric vehicles per one high power charging position in selected countries in Turkey in 2018 (Source: Own base on [22])

average number of plug-in EVs per one normal power charging position is 15 PHEVs, whereas the average number of plug-in electric vehicles per one high power charging positions equals 60 PHEVs.

3 Electromobility in Private Transport

What currently constitutes the most serious barriers preventing consumers from buying an electric cart in Poland is the high price and general lack of charging infrastructure. They consider the possibility to charge an electric car at home or at work to be the major factor. Assuming Poland's average daily kilometrage of 25 km and the realistic range of electric cars ranging between 100 and 150 km, one would have to recharge the electric car every several days. However, for those who do not own a parking place or a garage, the only alternative is to use the poorly developed network of public stations.

It is still difficult to travel between large towns on account of the charging infrastructure deficiencies, since fuel companies have only just begun installing the first charging stations at motorway traveller service points (MOP) next to major junctions. In accordance with the said act [2], 6000 normal and 400 quick EV charging stations are assumed to be installed in Polish agglomerations by 2020. It is the combination of the public system with the service points to be installed by larger businesses and smaller private charging stations which should enable unconstrained and convenient use of roads all around Poland.

However, there are still numerous other challenges which must be tackled to ensure the infrastructural development, including optimum siting of service points, acquisition/lease of land to build stations, capabilities of the existing power grid, financing for construction of charging stations, as well as such limitations as long charging time (compared to car fuelling time) which translates into a significantly lower service capacity of such stations. Consequently, the limitations applicable to the transport network in terms of the EV charging capabilities are major factors which affect the siting of charging stations. Searching for an optimum location for electric car charging points/stations, one must take numerous aspects into considerations, including:

- electric car's service range (typically 100–150 km),
- electric car charging time (at least 30 min),
- population density of the area analysed,
- number of electric cars operated in the given area,
- availability of the power supply infrastructure,
- availability of parking places (for car charging purposes) and land ownership aspects,
- distance between charging points located in different agglomerations in order to enable convenient travelling between towns.

4 Analysis of Transport Accessibility of Electric Car Charging Points/Stations in the Conurbation of Upper Silesia and Dąbrowa Basin

At the end of 2018, there were 24 electric car charging stations in the Conurbation of Upper Silesia and Dąbrowa Basin, scattered around different towns. There were 7 public-access charging points in Katowice at the end of 2018, with 22 such facilities under construction. As declared by the Municipal Office in Katowice, further 115 publically available electric car charging points are about to be commissioned in the town by 2020 [23].

The scientific literature discusses many different approaches and methods used to solve the problem of siting of charging points or charging stations for electric vehicles. Some examples are provided in A. Lam et al. [24, 25], F. Guo et al. [26], Z. Liu et al. [27, 28], L. Jia et al. [29], E. Pashajavid et al. [30], T. Chen et al. [31], Z. Drezner [32], D. Mayfield [9], I. Frade et al. [33] and J. Jung et al. [34]. When choosing a location of charging points or charging stations, the methods used most commonly for this purpose include: multi-criteria methods, multi-criteria decision analysis (MCDA), mathematical programming, gravity methods, optimisation methods, multiple location methods, heuristic methods, spatial interaction methods, weighted checklist methods, covering methods, regression analysis methods, game theory methods, location-allocation methods and many others. The aforementioned methods were used for siting of the selected charging points and charging stations for electric vehicles functioning in the Conurbation of Upper Silesia and Dabrowa Basin. It should also be noted that, in many cases, the MCDA method was used (with a different number and type of criteria). The author's own study involved a transport accessibility analysis performed for the EV charging points and charging stations in operation in the Conurbation of Upper Silesia and Dabrowa Basin.

This kind of information can be useful for planning and operating reasons. The graphical representation of some of the results have been presented in Fig. 3. The relevant range is marked with a circle of 25 km in radius (red colour), 50 km in radius (green colour) and 100 km in radius (blue colour). The ranges were assumed on the basis of the available literature of the subject. The central part of the Conurbation of Upper Silesia and Dabrowa Basin is characterised by high accessibility of the charging points and charging stations for electric vehicles currently in operation. Moreover, having assumed larger service ranges (50 and 100 km in radius), one can observe a high level of transport accessibility in the central part of the Conurbation of Upper Silesia and Dąbrowa Basin. The fewest charging points are located on the fringes of the Conurbation of Upper Silesia and Dąbrowa Basin, which automatically translates into poor accessibility of these areas by the means of transport in question.

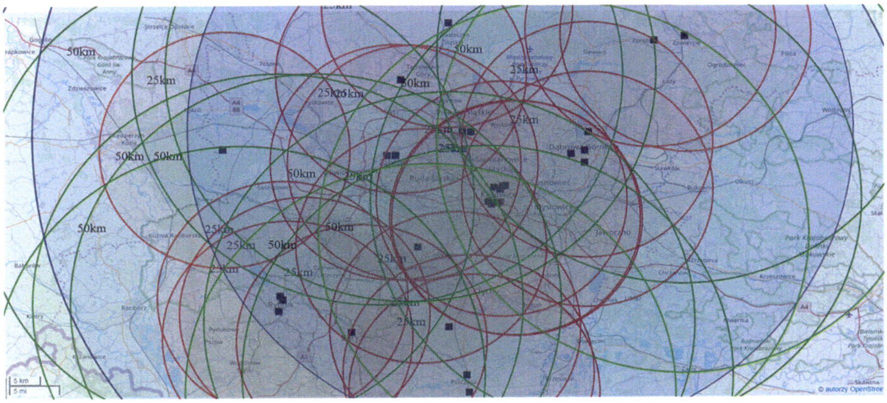

Fig. 3 Locations of the charging points and charging stations for the electric vehicles currently in operation on Upper—Silesian—Zagłębiowska Conurbation (R = 25, 50 and 100 km) (Source: Own research)

5 Electromobility in Public Transport

In global terms, a fair share of public transport has long relied on electricity, just to mention trams, trolleybuses, trains or metro services. In some Polish towns, the act on electromobility has induced local authorities to start replacing bus fleets with more ecofriendly vehicles or fully electric ones (including in Warsaw and Krakow). As an outcome of the collaboration between the central government and local authorities, a letter of intent was signed in 2017, with the municipalities undertaking to purchase 780 electric buses and 481 electric cars. The size of the electric bus fleet and the number of buses in general in selected Polish towns in the first half of 2019 has been summarised in Table 1. The data provided in Table 1 imply that combustion engine buses are still predominant in public collective transport. The most numerous fleets of electric buses are operated in Warsaw, Krakow and Zielona Góra. They are all municipalities which attach great attention to the quality of collective transport services as well as air quality in the local policies they implement. Furthermore, the attention to the matters of public space triggers maximisation of quality and minimisation of nuisance of transport, particularly in city centres. Many other municipalities in Poland are also interested in the electromobility problem, having declared their willingness to conduct pilot studies or purchase electric buses.

The experience in operating electric buses municipalities and communes have acquired to date is generally positive. An electric bus can cover an average range of 150 km, and slow charging typically takes place at depots (e.g. in Warsaw, Sosnowiec) and bus terminuses (e.g. line 154 in Krakow). Fast charging, on the other hand, is provided using pantographs (e.g. in Jaworzno). On top of that, one should consider the related environmental aspects, such as elimination of exhausts, travelling comfort and lack of noise when an electric bus is used. Other advantages of this mode of public transport include low operating costs and fleet maintenance

Table 1 Size of the electric bus fleet and the number of buses in general in selected Polish towns in the first half of 2019 (Source: own materials based on statistical data obtained from the [35])

City	The number of combustion, hybrid and gas buses	The number of electric buses	The total number of buses in the city
Białystok	261	0 (planned purchase 25 buses in 2020)	261
Bydgoszcz	190	11	201
Czestochowa	122	0	122
Gdynia	266	6	272
Jaworzno	59	23	83
Kalisz	65	11	76
Kielce	206	25	231
Kraków	625	26	651
KZK GOPi MZKP Tarnowskie Góry	1858	10 (Katowice) 3 (Sosnowiec)	1871
Legnica	65	0	65
Lublin	285	21	306
Łódź	409	0	409
Olsztyn	185	0	185
Poznań	466	0 (planned purchase 21 buses in 2020)	466
Rzeszów	180	10 (planned purchase 90 buses)	190
Szczecin	421	0 (planned purchase 11 buses in 2020)	421
Tarnów	96	0	96
Toruń	140	0 (planned purchase 6 buses in 2025)	140
Warszawa	1704	20 (planned purchase 140 buses to 2020)	1724
Wrocław	421	0 (planned purchase 10 buses in 2020)	421
Zielona Góra	76	0 (planned purchase 76 buses)	76

causing little problem. Nevertheless, there is also a serious disadvantage, namely heating under winter conditions which significantly reduces the battery life.

6 Conclusions

Governments of many countries all over the world have been introducing regulations aimed to motivate consumers to purchase the EV using various incentives, like the subsidy scheme which has already been deployed in 17 European countries (e.g. in Norway, Germany, France, Spain, Romania, Slovenia and the United Kingdom). In terms of the state support the EV buyers receive, Norway is an unquestioned leader. In 2017, electric cars accounted for 42% of all newly registered cars in Norway. Persons who buy the EV in Norway are—among other reliefs—offered exemption from 25% VAT charged upon purchasing an electric car (paid by those who buy combustion engine vehicles) as well as from administrative fees due upon vehicle registration, and they are eligible for free use of toll motorways. Moreover, owners of the EV are obliged to pay road tax at a reduced rate, while in some regions of the country they are not charged for the use of parking spaces and are entitled to use bus lanes. Plans made by governments of other countries also contribute to the popularisation of the EV, as some of them have announced a ban on sales of combustion engine cars (starting from 2025 in Norway, from 2030 in India, and from 2040 in France and the United Kingdom). According to forecasts, one can also expect prices of lithium-ion batteries to decline in the incoming years, and so will the EV purchase prices. The network of charging stations will keep developing. According to some sources, as many as 530,000,000 electric cars may be put into operation by 2040 worldwide.

As forecasted by the Alternative Fuels Market Observatory [36], prices of electric cars in Poland will start to gradually level up with those of traditionally propelled cars in the years 2026–2029. It is assumed to be affected by the technological progress, mainly including the decline in battery prices and the effect of production scale which is expected to translate into more affordable prices offered by automotive companies. This will coincide with a demand increase, enabled by a more highly developed charging infrastructure and fostered through public aid schemes. With reference to the analysis provided in the article, one can claim that, compared to other European countries, Poland is still at an early stage of electromobility development. The act on electromobility and alternative fuels obligates municipalities and communes to build electric car charging stations, but prices of electric cars are still disproportionately high compared to average remuneration of the Poles. There are no efficient incentive programmes that could foster purchasing an electric car. In spite of the foregoing, the growing environmental awareness of the society triggers interest in electric cars, both among private consumers and in the public collective transport sector. As provided in the article, results of the transport accessibility analysis imply that the central part of the Conurbation of Upper Silesia and Dąbrowa Basin is characterised by appropriate distribution of the existing charging points and charging stations for electric vehicles. Moreover, having assumed larger service ranges (50 and 100 km in radius), one can observe a high level of transport accessibility in the central part of the Conurbation of Upper Silesia and Dąbrowa Basin. The fewest charging points are located on the fringes of the Conurbation of Upper Silesia and Dąbrowa Basin, which automatically translates into poor accessibility of these areas by the means of transport in question. Nevertheless, it should be noted that the share of electric cars in the automotive market, so in the Conurbation of Upper Silesia and Dąbrowa Basin as in the whole of Poland, is still negligible. However, as the number of users of electric cars grows, it will be increasingly necessary to expand the EV charging infrastructure on account of the limited service capacity of the existing stations. It is now a greater challenge to travel between towns and agglomerations than to me within a conurbation, since—as the available data imply—the number of charging stations which can be used in interconnecting roads is largely insufficient.

Acknowledgements The present research has been financed from the means of the National Centre for Research and Development as a part of the international project within the scope of ERA-NET CoFund Electric Mobility Europe Programme "Electric travelling—platform to support the implementation of electromobility in Smart Cities based on ICT applications".

References

1. Parlament Europejski: Dyrektywa Parlamentu Europejskiego i Rady 2014/94/UE z dnia 22 października 2014 r. w sprawie rozwoju infrastruktury paliw alternatywnych. https://publications.europa.eu/pl/publication-detail/-/publication/d414289b-5e6b-11e4-9cbe-01aa75ed71a1
2. Kancelaria Sejmu: Ustawa z dnia 11 stycznia 2018 r. o elektromobilności i paliwach alternatywnych (Dz.U.2018 poz. 317), Warsaw (2018)
3. Gis, W., Kordel, Z., Menes, M.: Cars with electric drive and external costs of road transport. Przegląd Elektryczny. **10**, 287–292 (2013)
4. Jacyna, M., Lewczuk, K., Szczepanski, E., Gołębiowski, P., Jachimowski, R., Kłodawski, M., Pyza, D., Sivets, O., Wasiak, M., Żak, J., Jacyna-Golda, I.: Effectiveness of national transport system according to costs of emission of pollutants. In: Nowakowski, T., Mlynczak, M., Jodejko-Pietruczuk, A., Werbinska Wojciechowska, S. (eds.) Safety and Reliability: Methodology and Applications. Proceedings of the European Safety and Reliability Conference (ESREL), pp. 559–567. ESREL, Wrocław (2015)
5. Jacyna, M., Wasiak, M., Lewczuk, K., Kłodawski, M.: Simulation model of transport system of Poland as a tool for developing sustainable transport. Archiv. Transp. **31**(3), 23–35 (2014)
6. Kuczyński, W., Denis, A.: Electromobility as one of the ways to a low emission transport system. J. Mechan. Energ. Eng. **2**(1), 51–58 (2018)
7. Macioszek, E.: Electric vehicles—problems and issues. In: Sierpiński, G. (ed.) Smart and Green Solutions for Transport Systems. Advances in Intelligent Systems and Computing vol. 1091, pp. 169–183, Cham. Springer (2020)
8. Macioszek, E.: Freight transport planners as information elements in the last mile logistics. In: Sierpiński, G. (ed.) Integration as Solution for Advanced Smart Urban Transport Systems, Advances in Intelligent Systems and Computing, vol. 844, pp. 242–251, Cham. Springer (2019)

9. Macioszek, E.: First and last mile delivery—problems and issues. In: Sierpiński, G. (ed.) Advanced Solutions of Transport Systems for Growing Mobility, Advances in Intelligent Systems and Computing, vol. 631, pp. 147–154. Springer, Cham (2018)
10. Murawski, J., Szczepański, E.: Perspektywy dla rozwoju elektromobilności w Polsce. Logistyka - nauka. **4**, 2249–2258 (2014)
11. Sierpiński, G., Staniek, M.: Platform to support the implementation of electromobility in smart cities based in ICT applications—concept for an electric travelling project. Sci. J. Siles. Univ. Technol. Series Transp. **100**, 181–189 (2018)
12. Zawieska, J., Pieriegud, J.: Smart city as a tool for sustainable mobility and transport decarbonisation. Transport Policy. **63**, 39–50 (2018)
13. Macioszek E.: E-mobility infrastructure in the Górnośląsko - Zagłębiowska Metropolis, Poland, and potential for development. In: Proceedings of the 5th World Congress on New Technologies (NewTech'19), pp. ICERT 108-1-ICERT 108-4. NewTech, Lisbon (2019)
14. Decker, B., Hecimovic, H., Wołek, M.: Sustainable urban mobility planning in central eastern Europe: case examples from Poland and Croatia. Procedia - Social and Behavioral Sciences. **48**, 2748–2757 (2012)
15. Kadłubek, M.: Examples of sustainable development in the area of transport. Proced. Econom. Finance. **27**, 494–500 (2015)
16. Krawiec, S., Łazarz, B., Markusik, S., Karoń, G., Sierpiński, G., Krawiec, K.: Urban public transport with use of electric buses—development tendencies. Transp. Prob. **11**(4), 127–137 (2016)
17. Lejda, K., Mądziel, M., Siedlecka, S., Zielińska, E.: The future of public transport in light of solutions for sustainable transport development. Sci. J. Siles. Univ. Technol. Series Transp. **95**, 97–108 (2017)
18. Macioszek, E.: Conditions of oversize cargo transport. Sci. J. Siles. Univ. Technol. Series Transp. **102**, 109–117 (2019)
19. Macioszek, E.: Changes in values of traffic volume—case study based on general traffic measurements in Opolskie Voivodeship (Poland). In: Macioszek, E., Sierpiński, G. (eds.) Directions of Development of Transport Networks and Traffic Engineering. Lecture Notes in Networks and Systems 51, pp. 66–76. Springer, Chem (2019)
20. United Nations Conference on Climate Change COP 21. November 30th to December 11th, Paris (2015). https://unfccc.int/resource/docs/2015/cop21/eng/l09r01.pdf
21. Global EV: Outlook 2017. https://www.iea.org/publications/freepublications/publication/GlobalEVOutlook2017.pdf
22. European Alternative Fuels Observatory. www.eafo.eu
23. Witryna internetowa. https://www.bankier.pl/moto/katowice-115-punktow-ladowania-aut-elektrycznych-do-2020-r-311/
24. Lam, A.Y.S., Leung, Y.W., Chu, X.: Electric vehicle charging station placement: formulation, complexity and solutions. IEEE Trans. Smart Grid. **5**(6), 2846–2856 (2014)
25. Lam, A.Y.S., Leung, Y.W., Chu, X.: Electric vehicle charging station placement. IEEE Trans. Power Del. **25**, 510–515 (2013)
26. Guo, F., Inoa, E., Choi, W., Wang, J.: Study on global optimization and control strategy development for a PHEV charging facility. IEEE Trans. Veh. Technol. **61**(6), 2431–2441 (2012)
27. Liu, Z., Wen, F., Ledwich, G.: Optimal planning of electric-vehicle charging stations in distribution systems. IEEE Trans. Power Del. **28**(1), 102–110 (2013)
28. Liu, Z., Zhang, W., Ji, X., Ke, L.: Optimal planning of charging station for electric vehicle based on particle swarm optimization. Proc. IEEE Innov. Smart Grid Technol. Asia, pp. 1–5 (2012)
29. Jia, L., Hu, Z., Song, Y., Luo, Z.: Optimal siting and sizing of electric vehicle charging stations. Proceedings of IEEE Int. Elect. Veh. Conf., pp. 1–6 (2012)
30. Pashajavid, E., Golkar, M.A.: Optimal placement and sizing of plug in electric vehicles charging stations within distribution networks with high penetration of photovoltaic panels. J. Renew. Sustain. Energ. **5**(5), 101–118 (2013)

31. Chen, T.D., Kockelman, K.M., Khan, M.: The electric vehicle charging station location problem: a parking-based assignment method for Seattle. In: Proceedings of 31st USA EE/IAEE North America Conference, pp. 1–14 (2012).
32. Drezner, Z.: Facility location: applications and theory. Springer-Verlag, Berlin (2010)
33. Frade, I., Ribeiro, A., Goncalves, G., Antunes, A.P.: Optimal location of charging stations for electric vehicles in a neighborhood in Lisbon, Portugal. J. Transp. Res. Board. **2252**, 91–98 (2011)
34. Jung, J., Chow, J.Y.J., Jayakrishnan, R., Park, J.Y.: Stochastic dynamic itinerary interception refueling location problem with queue delay for electric taxi charging stations. Transp. Res. Part C Emerg. Technol. **40**, 123–142 (2014)
35. Zespół Doradców Gospodarczych TOR: Elektromobilność w Polsce. Perspektywy rozwoju, szanse i zagrożenia. Komunikator PR, Warszawa (2017)
36. Obserwatorium Rynku Paliw Alternatywnych. http://www.orpa.pl/

Wireless Sensor Networks for Traffic Monitoring

Wiktoria Loga, Justyna Sordyl, and Artur Ryguła

Abstract Nowadays, one of the most dynamically developing issues inherent to sustainable development is a Smart City concept. Intelligent Transportation Systems become a tool for providing information on traffic conditions as well as traffic control, therefore they require modern and effective measurement methods. Wireless Sensor Networks (WSN) are an example of those solutions as they (due to their functionality) enable continuous and low-cost road network monitoring in selected area.

Authors of the research used Wi-Fi based WSN for traffic streams detection and made an attempt to evaluate the network performance. The presented measurement results were obtained using devices and systems integrated by ITS testing ground located in University of Bielsko-Biala. The research is a continuation of a previous work regarding wireless multisensor networks suitability for traffic flow detection and tracking.

Keywords Wireless sensor network · Wi-Fi detection · Traffic flow tracking · Pedestrian flow

1 Introduction

Along with the road traffic volume increase, the competent traffic flow control becomes a challenge for infrastructure administrators. The actual and reliable data coming from the road net and their analysis and characteristics are the valuable tool for the road traffic service and the future investment planning. The information demand is especially essential at urbanized areas where the traffic flow is higher and transport congestions, bottlenecks, increasing noise and air pollution occur more often.

W. Loga (✉) · J. Sordyl · A. Ryguła
University of Bielsko-Biala, Bielsko-Biala, Poland
e-mail: wloga@ath.bielsko.pl; jsordyl@ath.bielsko.pl; arygula@ath.bielsko.pl

M. Suchanek (ed.), *Transport Development Challenges in the 21st Century*, Springer Proceedings in Business and Economics,
https://doi.org/10.1007/978-3-030-50010-8_14

For the road traffic engineering, the information on the motor and pedestrian traffic flow is one of the most essential among the basic characteristics describing the traffic. The road traffic detection and supervision are also the fundamental elements of functioning the Intelligent Transport Systems and development of the Smart City concept. This concept assumes use of digital information and communication networks to integrate the city systems in order to improve their efficiency and interactivity [1, 2]. Use of these elements allows to meet the increasing requirements of the urbanized area habitants as regards fulfillment of their transport needs.

2 Detection Systems

In view of high demand, the detection system market has been developing very dynamically for the recent years. Besides the detection itself, the modern systems allow to obtain data concerning the traffic distribution, the speed, the traffic density (headway measurement), the individual vehicle classification and weight. Multitude of solutions and technologies provides a variety of products. The presently used solutions are based on detection of effects accompanying the passing vehicle. Actually used measurement technologies can be divided into two basic groups: based on invasive and noninvasive sensors.

The first group includes solutions where detectors are mounted directly in the pavement with disarrangement of the roadway structure. This group contains inductive loops, magnetometers, microloop probes, pneumatic road tubes, piezoelectric cables, or other alternative weigh-in-motion sensors [3]. The invasive systems are characterized by high accuracy of measurement, however their installation and maintenance afford necessity of the road traffic interruptions. In addition, intervention in the pavement structure leads to faster exhaustion of its fatigue life and the maintenance work can result in necessity of replacement or recalibration of the sensors installed.

The second group includes cost-effective, noninvasive solutions; the sensors are usually installed above the traffic lane or at the roadside. This group contains, among others, video image processing, microwave and laser radars, passive infrared, ultrasonic, passive acoustic array, sets of diverse sensing technologies such as passive infrared Doppler radar. Some of the noninvasive technologies can reach the detection accuracy comparable to the invasive solutions. At present, the very popular solutions use also the low-cost and wireless sensor networks. The assembly and maintenance work influence on the road traffic is limited up to minimum. The usual mobility of these solutions allows also easy change of detector locations, e.g., the solutions based on detection of activity of devices equipped with Wi-Fi or Bluetooth module [4, 5]. In presented research authors used Wi-Fi based WSN for traffic streams detection and made an attempt to evaluate the network performance.

3 Technology Used in the Study

The wireless sensor networks (WSNs) present a set of any number of compact, low-cost devices (sensor nodes) that intercommunicate wirelessly and their task consists in event detection and evaluation of its parameters. A typical sensor node consists of a sensor, a processor, a memory card, a communications module and a power supply source. The network nodes exchange data with the base station called the sink node that is used to collect and process data in the centralized mode; it performs also a role of an interface between the detection area and the user. A typical WSN architecture is shown in the Fig. 1.

In order to obtain proper functioning the Wireless Sensor Network, its competitiveness in relation to other solutions and its measurable advantages, it is necessary to meet the following requirements [6]:

- Scalability—ability for a network to grow in terms of the number of nodes attached to the wireless sensor network.
- Responsiveness—ability of the network to quickly adapt itself to change in the topology.
- Reliability— probability that a system will perform satisfactorily during e.g., node failure until the other nodes collecting similar data in the same area are accessible.
- Increased range—many wireless sensors can replace one wired item to cover the larger region.
- Mobility—ability to handle mobile nodes and changeable data paths.

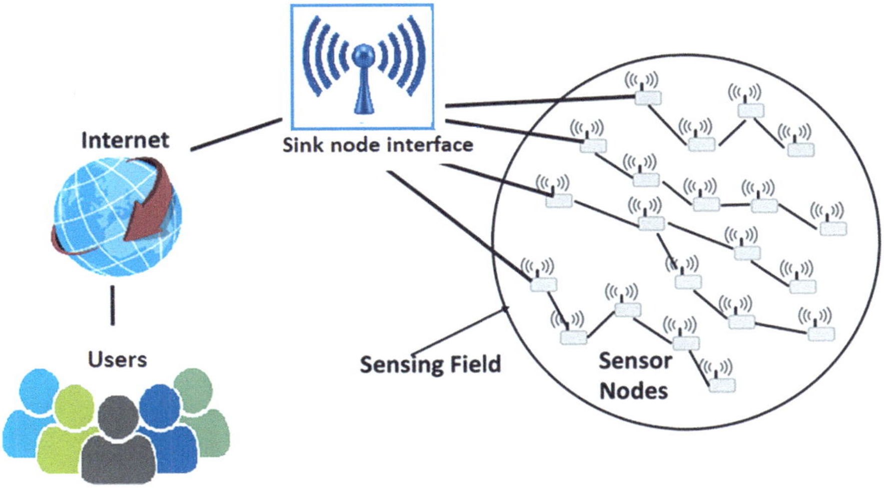

Fig. 1 Wireless sensor networks architecture (Source: own study)

Application of this solution in transport can be used to for real-time data detection, processing, and transfer. They are necessary for the optimum and complex road traffic management and adaptation to a dynamically changing situation in the detection area. WSN allows, among others, data collection concerning the location, the road traffic conditions, the route and duration of passage/drive, the object speed and acceleration, weather conditions, etc. [7].

4 ITS Testing Ground

In 2016, at the campus of University of Bielsko-Biala, the instructive and research ITS testing ground was built to carry out the road traffic studies. The testing ground is divided into the monitoring center that integrates all subsystems and the measuring apparatus implemented in the road system.

The measuring part (Fig. 2) includes the following elements of the road infrastructure [8]:

- Table of variable content of RGB type,
- 3D scanner allowing the vehicle outline classification and detection,
- Scales to preselect vehicles in motion in the WIM (Weigh In Motion) mode,
- Inductive loop set to classify vehicles and specify their velocity,
- Weather station,
- Overview camera set,
- Bluetooth activity sensor system.

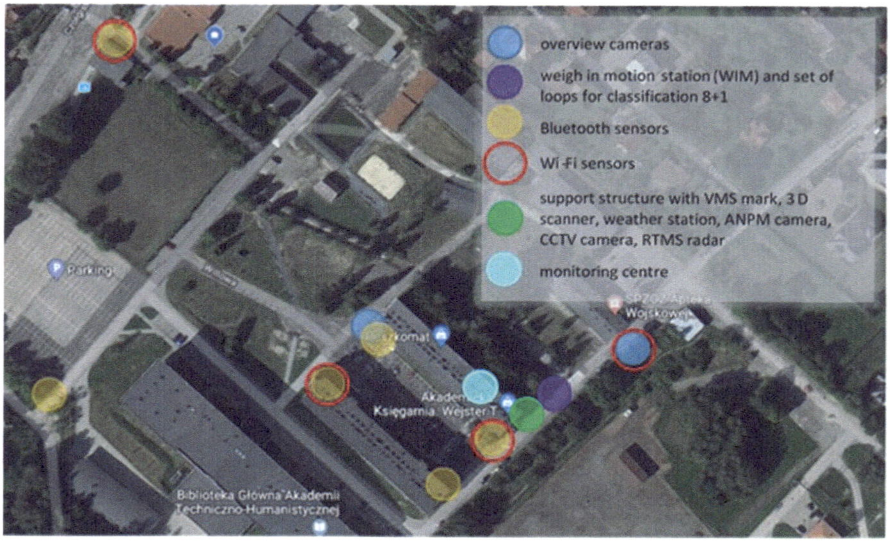

Fig. 2 Location of ITS testing ground elements (Source: own study)

5 Using WSNs on the Campus Area

Within the work at the testing ground, the traffic stream detection is developed, especially in the area of estimation of the vehicle and pedestrian stream rates and their demarcation. So far, the studies concerning, among others, effectiveness and specificity of the Bluetooth detection system operation were carried out with reference to the data coming from the remaining integrated infrastructure of the testing ground [9]. It results from the research being carried out that the system based on Bluetooth detection, implemented on the municipal communications network, is able to detect between 10% and 35% of all vehicles [8, 10].

To extend the research and instructive capabilities, the testing ground was developed in 2019 with a new functionality—Wi-Fi activity sensor system. In this way, the comprehensive research on determination and demarcation of the vehicle and pedestrian streams has been made possible. Principle of sensor network operation consists in search of Wi-Fi wireless radio activity. Each sensor (Fig. 3) listens to 802.11 traffic and detects all mobile devices in active Wi-Fi scanning mode. Each device with the active Wi-Fi module is registered in the database together with its ID (MAC address) and the accurate time marker, when it occurred in the detection range.

Moreover, the individual hardware address of MAC network card of each device provides information on the given unit origin. The address (Fig. 4) is the 48-bit number in the hexadecimal format; the first 24 bits of the number mean the network card manufacturer, the remaining 24 bits are the unique item identifier assigned by its producer. At present, the user privacy protection trend gains popularity and thereby many manufacturers of mobile devices use MAC-Anonymity. This measure purposes randomization of MAC address by use of its encryption by means of variable and random character set. Consequently, the unique identification of the specific device and its tracking is not possible within the longer period of time.

In connection with the above, the scheme of research carried out using the Bluetooth technology has been repeated using the new technology. This allowed to test and check the Wi-Fi sensor usefulness for estimation of the pedestrian stream

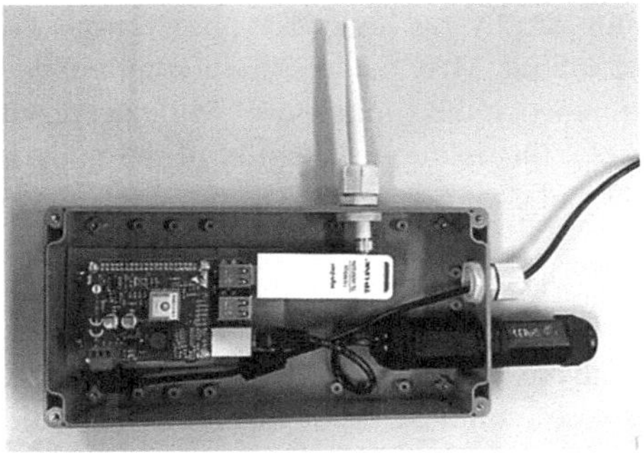

The sensor consists of:

- IEEE 802.11 Wireless USB Adapter,
- 4 dBi external antenna,
- Raspberry Pi microcomputer,
- power cord and Ethernet cable,
- sensor enclosure.

Fig. 3 Wi-Fi sensor and its elements (Source: own study)

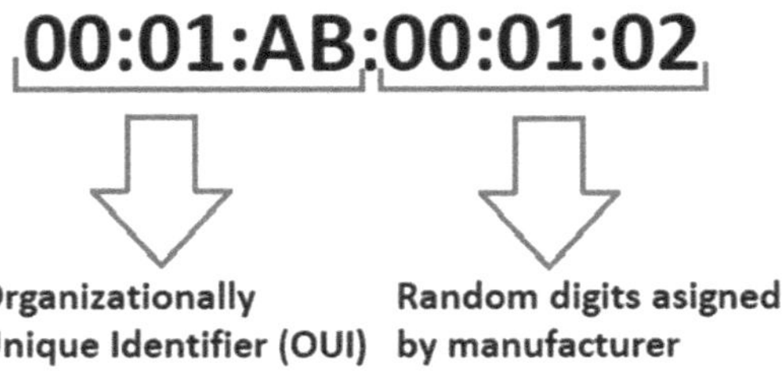

Fig. 4 MAC address format (Source: own study)

parameters. As the previous research of the wireless sensor networks based on Wi-Fi detection has shown [11], there is correlation between the pedestrian streams and the number of the detected Wi-Fi devices. Thereby, the system usability in pedestrian traffic detection has been confirmed and, consequently, complementarity in relation to the Bluetooth WSN detection technology implemented before has been ascertained.

6 Measurement Data Analysis

6.1 Analysis of Number and Type of the Detected Devices

To specify the detection capability of the used sensor network, the analysis was carried out based on the data collected within the week (08/04/2019 to 14/04/2019) for the nodes A and D (Fig. 5). These are the sensors located at the entrance and the main exit from the campus site. It shall be mentioned that the outpatient clinic, the main University administration building and the measurement infrastructure of the ITS testing ground are located in the vicinity of the node A.

Effect of the mentioned devices is apparent in the Fig. 6 where, for the node A, the number of the unique MAC addresses in comparison to all detections decreased on average by about 43% for business days while by 33% on weekends. This difference can result from lack of office devices activity on weekends that is dictated by the university administration worktime. The differences in the number of the devices detected before and after data cleanup are by far lower for observations coming from the node D. The number of the unique MAC addresses in comparison to all detections during the whole week remained at the similar level while decreasing on average by about 26%. The most likely, this is due to location of the node D that does not sit next to the organizational and administrative facilities or the advanced measuring apparatus.

To verify the generic structure of devices detected by the wireless sensor network, the authors made an attempt to identify manufacturers of the detected devices equipped with the Wi-Fi module. Analysis of first member of the MAC address allowed identification of manufacturers of some of the devices. The identification was carried out using the official database of The IEEE Registration Authority. It should be emphasized that due to the data randomization process mentioned in the

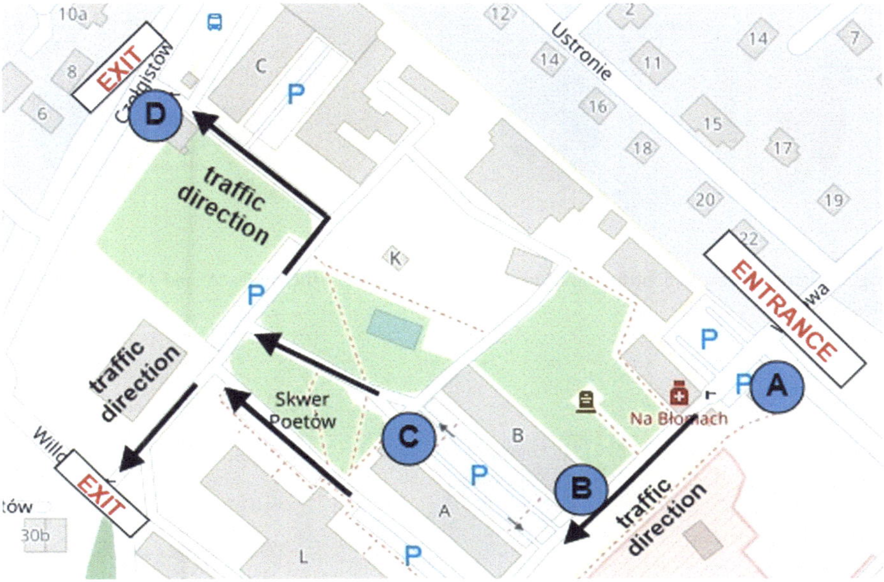

Fig. 5 Arrangement of Wi-Fi network nodes and vehicle traffic organization at the campus site (Source: own study)

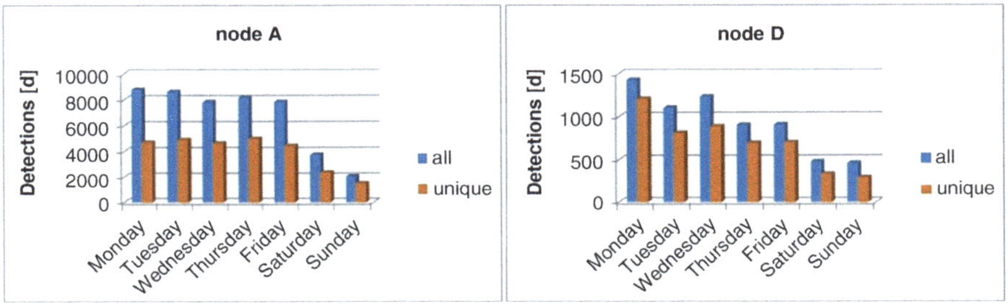

Fig. 6 Comparison of the detection values before and after data preliminary filtration for the nodes A and D (Source: own study)

Chap. 5, the large part of the detected MAC addresses started with the universal prefix DA:A1:19 belonging to Google. In view of MAC-Anonymity, any detected object could be identified as a mobile device, however randomization precluded to specify a manufacturer precisely. All observations have been cleared from interferences through separation, pure and simple, of unique MAC addresses. The Fig. 7 shows comparison of structure of the devices detected by the chosen network nodes. In case of the node A, the identified MAC addresses made 75%, i.e., they represented the mobile devices definitely. The unidentified 25% of observations can represent both mobile devices with randomized MAC address using the variable prefix and office devices at the administrative facilities or ITS testing ground mea-

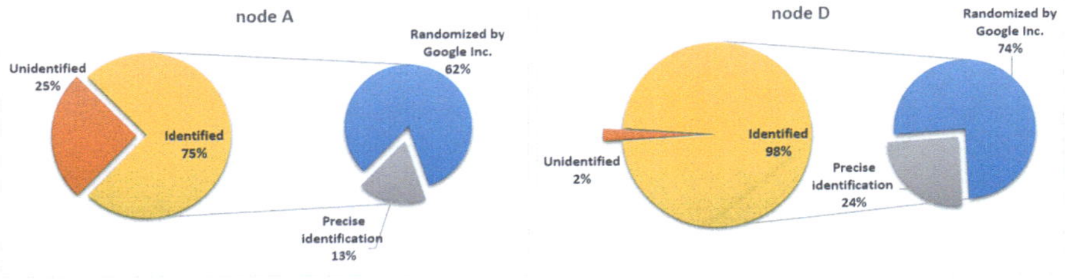

Fig. 7 Percentage of identified and unidentified devices for the nodes A and D (Source: own study)

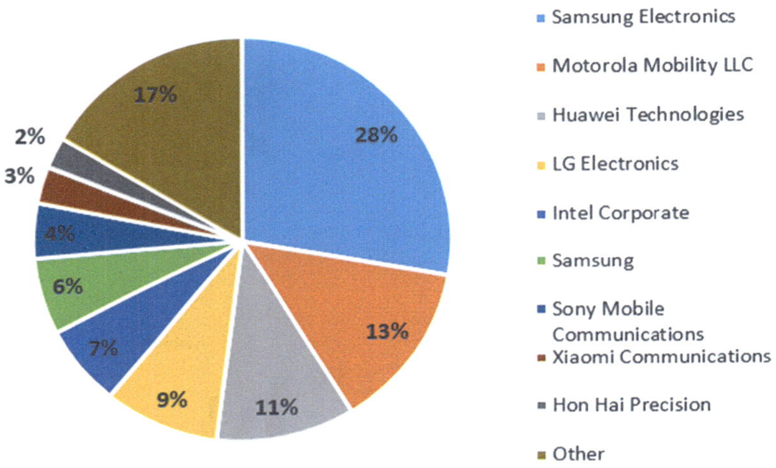

Fig. 8 Participation of device manufacturers identified precisely from the node A (Source: own study)

suring apparatus. In case of the node D, the unidentified MAC addresses made only 2%. This confirms the earlier conjectures that there are no interferences coming from devices that are not mobile at the location under examination. Consequently, 98% of devices could be identified as the mobile devices, whereout 74% were classified as devices with Android system or iOS with randomized address and 24% were attributed explicitly to the concrete manufacturer.

Among all devices detected by the node A, 62% could be classified as mobile Android or iOS devices (judging by randomized Google prefix), however 13% of devices could be identified precisely and attributed to the manufacturer (Fig. 8).

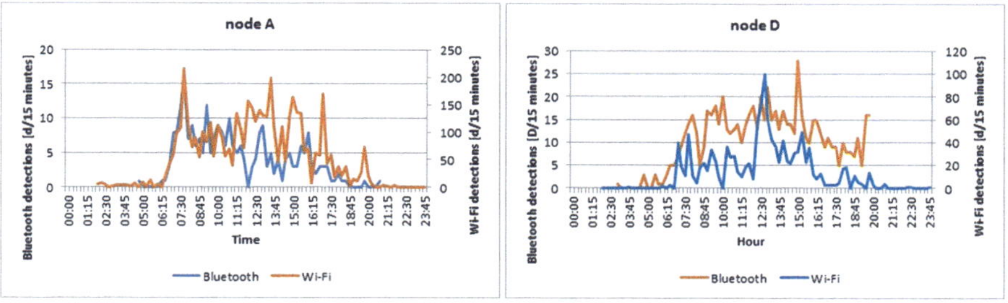

Fig. 9 Distribution of the detected MAC addresses for the Wi-Fi and Bluetooth wireless communications systems (Source: own study)

6.2 Comparative Analyses of WSN Detecting Wi-Fi Activity and the Network Detecting Bluetooth Activity

To compare numbers of Wi-Fi device and Bluetooth device detections, the detected unique MAC addresses have been listed for the chosen measuring day (08/04/2019). The analysis included data from the nodes A and D located at the entrance and the main exit from the campus site, respectively (Fig. 5). The hourly distribution of number of the detected MAC addresses for the individual wireless communications systems is shown in Fig. 9.

For the node A, there is the noticeable morning rush hour about 7:45 a.m. registered for the detected devices, both Wi-Fi and Bluetooth; this corresponds to the time of work and course commencement at the University. It shall be noticed that the number of Wi-Fi device detections is over 10 times higher than Bluetooth device detections. At the exit node D, the main rush hour during the day happens on 12:45 for Wi-Fi detection and on 15:00 for Bluetooth detection. This difference can be caused by the higher detectability of pedestrian streams using Wi-Fi detection and of vehicle streams using Bluetooth detection. These hours correspond to the typical completion time of course for students (about 13:00) that, in large part, walk on foot due to limited possibility to enter the campus with vehicles. However, for the network based on Bluetooth technology that detects mainly vehicle streams, the rush hour was noted down about 15:00 what corresponds to the work completion time of the University administration. In case of the exit node D, the difference in number of detections between the described technologies is by far lower, i.e., they differ only four times. This can result from the fact that the exit at the node D is not the only exit for pedestrian streams since the internal vehicle traffic organization at the campus is not binding for pedestrians.

7 Conclusion

The purpose of the research was using Wi-Fi-based WSN for traffic streams detection and the evaluation of the network performance. The analyses shown in the article are the continuation of research on wireless sensor network usefulness for detection of vehicle and pedestrian streams. All of the analyses shown are of the preliminary research nature in view of the relatively new implementation. The research carried out before by the authors confirmed explicitly the legitimacy to use the networks based on Wi-Fi and Bluetooth activity detection to detect vehicle and pedestrian streams. The analyses presented in the paper purposed the further research of the implemented Wi-Fi activity sensor network as regards the system functioning efficiency and stability.

While comparing the findings of research carried out within the week (08/04/2019 to 14/04/2019) with the results for the chosen measuring day (18/01/2019) shown in the previous research [11], the high repeatability of indications can be found. This is confirmed by stable sensor operation and almost identical level of type identification of the devices recognized.

A great influence of peripheral equipment of the administrative facilities and the instructive and research test ground on the node A detection quality has been shown; the equipment communicated the node continuously while disturbing the number of devices detected in fact. This influence diminishes considerably on weekends when there is lower number of interference sources. This fact is confirmed by the data obtained from the node D located far away from the administration buildings and the measuring apparatus of ITS test ground. They show smaller differences between the raw and cleared data and the very high level of the detected unit identification including the mobile device participation of 98%.

In the comparative analysis of two WSNs functioning at the University site, the similar trend was observed in detections at the campus entrance—the highest rush hour was noted down at 7:45 a.m. When considering the data from the nodes located at the exit, the discrepancy was observed for the time of afternoon rush hour. This situation is connected with different times of the courses and the administration worktime. The difference between the data obtained from the network detecting Wi-Fi activity and the Bluetooth network results from the detected stream types. The Wi-Fi WSN detecting mainly the pedestrian streams [11] has shown the rush hour about 13:00 and this is connected with the limited possibility for students to enter the campus with vehicles and the fact that they walk on foot mainly at its site. However, for the network based on Bluetooth technology that detects mainly vehicle streams, the rush hour was noted down about 15:00 what corresponds to the work completion time of the University administration.

References

1. Azkuna, I. (ed.): Smart Cities Study: International Study on the Situation of ICT, Innovation and Knowledge in Cities. The Committee of Digital and Knowledge-based Cities of UCLG, Bilbao (2012)
2. Giffinger, R., Fertner, C., Kramar, H., Kalasek, R., Pichler-Milanoviü, N., Meijers, E.: Smart Cities: Ranking of European Medium-Sized Cities, Centre of Regional Science (SRF), p. 11. Vienna University of Technology, Vienna (2007)

3. Mimbela, L.E., Klein, L.A.: Joint Program Office for Intelligent Transportation Systems, Washington DC. (2000). https://www.fhwa.dot.gov/ohim/tvtw/vdstits.pdf
4. Fernández-Lozano, J., Martín-Guzmán, M., Martín-Ávila, J., Gar-cía-Cerezo, A.: A wireless sensor network for urban traffic characterization and trend monitoring. Sensors. **15**(10), (2015)
5. Fernandez, A.A., Gajda, D.: Comparing Wireless Traffic Tracking with Regular Traffic Control Systems for the Detection of Congestions in Streets, "Smart Cities", pp. 42–51. Springer, Cham (2016)
6. Rhee, S., Liu, S.: Maximizing Data Reliability in Wireless Sensor Networks. Sensor Magazine Online July 2005. (2005). http://www.sensorsmag.com/articles/0705/23/main.shtml
7. Kafi, M.A., Challal, Y., Djenouri, D., Doudou, M., Bouabdallah, A., Badache, N.: A study of wireless sensor networks for urban traffic monitoring: applications and architectures. Proc. Comput. Sci. **19**, 617–626 (2013)
8. Grabara, A., Płosa, J., Ryguła, A.: The evaluation of the efficiency of determining traffic volume using bluetooth systems. In: Transport Problems' 2015: VII International Scientific Conference, IV International Symposium of Young Researchers: Proceedings, pp. 149–156 (2015)
9. Ryguła, A., Loga, W.: Wykorzystanie systemu Bluetooh do wykrywania strumieni pojazdów i pieszych, Autobusy, part 18, no. 12 (2017)
10. Konior, A., Brzozowski, K., Maczyński, A., Ryguła, A.: A concept of extension of the OnDynamic system with module for monitoring road traffic impact on the urban environment. Arch. Transp. Syst. Telemat. part 9, no. 2 (2016)
11. Loga, W., Ryguła, A., Sordyl, J.: Using wireless sensor networks for vehicles and pedestrian movement tracking. In: Paper Presented at Conference Transport Systems Telematics '19, 27.02.2019–02.03.2019, Jaworze (2019)

Electronic Fare Collection Systems vs. Passenger Rights and Effectiveness of Services Provision

Grzegorz Dydkowski

Abstract The chapter presents premises and benefits resulting from e-ticket systems implementation, in which the obligation was adopted to register each entry when travelling. The check-in registration system may be used to study passenger flows and this in turn to optimise the transport offer on a current basis and to carry out settlements with entities financing the urban public transport from public funds. Also analysis and assessment were performed, to what extent the registration obligation is permissible in the existing Polish legislation and whether it violated the collective consumer interest and is inconsistent with decency. This issue is related to dynamic development of IT technologies and the origination and use of all sorts of IT solutions. It enforces the updating of standards and adapting them to the changing social and economic environment.

Keywords Electronic fare collection systems · Public transport · Passenger rights · Urban transport

1 Introduction

The implementation of fare collection systems utilising e-tickets with registration of entering the vehicle (check-in) in the case of bigger cities and numerous operators is a relatively big and complicated investment project. It requires creating an appropriate infrastructure in vehicles of the urban public transport and a sales network, ensuring data transmission and processing, and their availability in an online standard, issuing a relevant number of cards as well as ensuring the entire system operation continuity. Also a change of hitherto rules of services use is important, where passengers use an electronic card instead of a paper ticket. During the system

G. Dydkowski (✉)
Department of Transport, University of Economics in Katowice, Katowice, Poland
e-mail: grzegorz.dydkowski@ue.katowice.pl

M. Suchanek (ed.), *Transport Development Challenges in the 21st Century*, Springer Proceedings in Business and Economics,
https://doi.org/10.1007/978-3-030-50010-8_15

designing, apart from changes of habits, it is also necessary to consider to what extent the assumptions comply with the binding legislation, in particular in the field of consumer rights and of personal data protection. The situation is complicated by the fact that many issues are based only on the interpretation of legal acts provisions, made most frequently only after the system implementation and start.

The application of entry (check-in) systems registration is often crucial from the point of view of economic benefits resulting from the implementation of fare collection systems based on electronic cards. Because it enables collecting the data on transport, which later on can be used to create a transport offer and in settlements related to the public financing of the public transport. So far the literature addresses the issues of fare collection systems functionality, standards used, and infrastructure necessary to implement such systems (P. Krukowski, W. Rokicki, K. Hebel, A. Urbanek, G. Dydkowski, B. Kos, J. Gnap, J. Paddington) and provides presentations of entities implementing such systems, like Asseco Poland S.A., Mennica Polska, R&G Plus Sp. z o.o., and PZI Taran Sp. z o.o. Research within the European Union is being conducted on a large scale [1]. Unfortunately, despite already quite common application of fare collection systems using e-cards, the legal issues or the carried out interpretations of the right to collect data based on electronic cards reading were not the subject of research and hence there is a lot of ambiguity. These issues are in the area of interest of the Office for Competition and Consumer Protection (UOKiK) and formerly of the General Inspector of Personal Data Protection—now the Personal Data Protection Office. The experience shows also that the aforementioned authorities approach the matter of e-tickets compliance with legislation on an individual basis, examining individual cases separately, and at the same time there are no systemic guidelines or so-called best practices in this field.

The chapter presents premises ad benefits resulting from e-ticket systems implementation, including systems of each entry registration when travelling. Against this background also the assessment has been presented, to what extent the fare collection system with the use of e-cards and with entry registration can violate the passenger rights. This issue has a specific economic dimension, which is difficult to overestimate. Because if the system of entry (check-in), and possibly exit (check-out), registration is given up, then at the same time the collection of data indispensable now to manage the transport offer and the system of tariffs and settlements of public funds allocated to the public transport are given up. The research was carried out based on the example of fare collection system using electronic cards, implemented by the Municipal Transport Union of the Upper-Silesian Industrial Region in Katowice [2]. This is a large system [3]—the cost of implementation including a 5-year period of system maintenance amounts to approx. PLN 180 million, and in the period between October 2015 and the end of 2018 approx. 500,000 cards have been issued. Correspondence with the Katowice Branch of the Office for Competition and Consumer Protection and with the Personal Data Protection Office has been carried out from 2016, finished with the Decision of the UOKiK President [4]. Now the judgement of the Regional Court in Warsaw is expected, where appeal against the UOKiK President's decision was filed [5].

2 Premises for Fare Collection Systems Using Electronic Cards Implementation in the Passenger Public Transport

The end of the twentieth century and a dozen or so of the twenty-first century years are marked by dynamic development of IT technologies and their deployment in transport. Advanced IT technologies, data collection and processing as well as big data sets management, the use of software with very developed functions, of computer networks, wireless communication, data transmission, and location systems are being developed, implemented and used in current operations. This expands capabilities of entities and even creates new conditions to carry out business and to manage the transport processes and also the entire area of services distribution, sales, and promotion—i.e. the front office. IT systems become an element combining into a whole individual stage of the transport system, enable quick access to the information and direct control of processes. The reduction of hardware prices, or maintaining the prices and offering hardware of greater capabilities, creates good conditions for the growing scope of performed functions, results in increasingly wide application of IT technologies, including in the urban transport.

The IT sector has developed and stabilised in Poland during the last twenty years. Deliveries of hardware and software, and also the deployment, are offered by big entities with substantial equity and financial resources and with many years of experience, as well as thousands of smaller entities flexible in operations, specialising in delivering and maintaining systems and modules in the field of selected areas of entity operations. A mature IT sector means also a greater availability of IT solutions and reduction of risk existing during IT systems implementation. The availability of IT technologies and areas of their application systematically grow, and entities—including in the urban transport—implement solutions, which are comprehensive, technologically advanced, and feature expanded functionality [6].

The progress in information transfer technologies has special importance in the transport business, including the public transport. Public transport services are not provided in one place, the possibility to transfer information facilitates the supervision and intervention in the event of disturbances, occurring on transport lines, frequently passing vast areas, outside dispatcher and control points. IT technologies ensure communication and information not only about the current location of vehicles, but also everything related to the fare collection systems.

Apart from location, the fare collection systems are an important area of IT systems implementation in the public transport. The type of fare collection for public transport services and the type of carrier on which the information about the collected fare is recorded are factors affecting the adopted tariff solutions, in particular possibilities to diversify prices, the flexibility of tariff changes, the used promotions and solutions in the field of tariff correspondence. The fares collection using electronic tickets and the registration of the number of rides enables using the data not only in relation to the fares charging, but also to the transport offer planning, to carrying out current changes and settlements, primarily with entities financing from public funds (in Poland municipalities) the urban public transport, as well as settlements between entities during common tariff projects [7].

Electronic cars provide potentially biggest possibilities to collect and process data about urban transport rides. In terms of the technologies applied—modern solutions are used. Chip (smart) cards are utilised, allowing to acquire data about rides and collected fares for the public transport services use. The introduction of fare collection systems based on electronic cards provides a possibility to acquire in a digital form the data related to the service use—it is easier to process the data and to use it to control the equipment, e.g. entries to transfer point areas. Electronic card technologies together with systems of legal and organisational solutions create a possibility to pay for services provided by various operators (not only related to the urban transport and vehicles parking) and to perform relevant related settlements. Therefore in a part of cases the implemented systems assume expansion of the services scope, where the card can be used e.g. in other municipal services. Solutions should be designed to provide a possibility of breaking down the payments made into individual services and into individual operators, which is especially important from the point of view of urban transport tariff integration.

In the case of electronic cards the data originating from fare collection systems can be the basic source of information about carried out rides and thereby about settlements between entities. It is possible to determine the number of passenger-kilometres (registration of check-ins and check-outs and of the distance covered by the vehicle between individual stops) or the number of passengers (only check-ins registration). The system may be supplemented also with data originating from the observation of the number of passengers directly in a vehicle or from the video monitoring. They can be applied in the transport offer planning processes and also for settlements related to the public financing of the public transport or to identify the amount of lost revenue resulting from the application of entitlements to concession and free travels.

The degree of IT deployment difficulty in the urban public transport depends on the solution scale. For small systems of urban public transport there are no major problems at the implementation of solutions in the field of automatic recording of the revenue amounts for individual lines or their sections. The fact that the implementation cases of fare charging based on e-cards, in the entry-exit or only entry identification system, is easiest in small and medium cities, could be an example. With the increasing city size, the number of operators, as well as with diversification of public transport reach and of the transport offer the project becomes more and more costly and complicated. Situations, where full functionality of fare collection systems using e-cards has not been achieved, most often do not result from technological conditions, but from organisational, legal or social reasons, or from mistakes made during a new solution promotion. The use of one electronic card in public transport systems, participated by many entities with various tariff regulations, requires relevant institutional solutions or using bank services for funds settlement.

The urban public transport is subject to regulation both in the field of price setting and of terms and conditions of transport services provision. Moreover, via the urban public transport price also the social, health protection, or related to the city land development policies are pursued within quite broad scope. Therefore it is possible to state that the obtaining of benefits depends in the possibility to implement and to accept new solutions in the field of prices, which to a large extent is beyond decisions of urban public transport managing units or operators [8].

It is also necessary to add that during the nearest years the IT technologies will become basic transport integration tools in the field of technical and technological solutions. So far electronic card systems are implemented primarily in urban transport systems or independently and separately in urban transport systems and in regional railway transport systems. Actions in favour of the tariff integration, but based on a common carrier, which will be an electronic card, are being undertaken now. Because cards valid in urban public transport systems are widespread, assumptions are made that it will be possible to use them also in the railway transport. Over time such systems should be tools used to a larger and larger extent during the transport offer planning, timetables coordination, traffic management, provision of information to passengers, and for settlements of funds between entities. They are also positively evaluated by the residents [9].

3 Passenger Rights in the Context of Forbidden Contract Clauses and of Personal Data Protection

3.1 The Obligation to Register the Vehicle Entering (Check-In) Each Time

The implementation of modern systems for public transport management is related to the acquisition and processing of a broad range of data. Such data are related not only to the location, parameters of means of transport traffic, and the size of transport performed by individual means of transport, but also necessarily to the location of persons—city card users and sections on which they were travelling. The right and legitimacy of such data collection by the public transport systems managers can raise doubts. In practice this means a question to what extent the organiser or carrier are authorised to order passengers, who have purchased and activated a season ticket without a limit of rides, to register each time the ride by carrying out the check-in, and possibly also carrying out the check-out. Together with systems development another question will arise, to what extent the organiser or carrier will be authorised to collect the data about passenger's ride on a specific route, in an automatic way, without the obligation to register the card at the entry, and possibly at the exit from the vehicle as well. In particular doubts related to each entry registration are articulated in the case of season tickets without a limit of rides, which have been previously bought and activated.

The right to impose obligation of registering each time the travel start results from the provision of Art. 16 of the Act of 15 November 1984 on Transport Law [10]: 'A transport contract is concluded by purchasing a travel ticket before the start of travelling or by meeting other conditions of access to a means of transport determined by the carrier or by a collective public transport organiser, and if they are not determined—by the fact of taking place in a means of transport.' That means that the legislator authorised public transport organisers to determine conditions of access to means of transport, which passengers must meet to conclude a transport contract. Such conditions are most frequently determined in the tariff or in another document regulating the rules of electronic cards use, where it is then possible to introduce the requirement of registration of entering the vehicle by holders of season tickets encoded on an electronic card or encoded entitlements to free travelling. In the case of establishing the obligation of registration, the fact of having an encoded season ticket on the card, without checking-in at each entry to a vehicle, does not meet the access conditions to means of transport determined by the organiser. At the same time attention should be drawn to the fact that the provision of Art. 16 of the Transport Law [10] gives a possibility to determine the access conditions to means of transport by organisers or carriers and does not limit or create a close catalogue of their advisability introduction, and does not require any particular justification. It is possible to assume that such conditions should only not violate general principles of social coexistence and obviously should not infringe other regulations by law. So in this context it does not seem that the obligation to register the card is an illogical and arduous action when entering a means of transport.

The opinion about a possibility to demand by an organiser or carrier to register each entry in the case of a season ticket purchased previously is confirmed by the reasons of the Voivodeship Administration Court in Białystok, which having examined at Division I during the hearing on 16 April 2014 of the case resulting from the complaint against the Resolution of Białystok City Council from April 2011, related to the determination of additional charges due to transport of persons, luggage or animals taken aboard a means of transport, and the amount of the manipulation charge, dismissed the case [11]. The reasons state that first it is necessary to determine, whether the term of missing appropriate transport document comprises non-registration at the entry to the vehicle of an electronic card with the encoded ticket. So it is necessary to clarify the term 'appropriate transport document'. Provisions of the new Transport Law do not comprise such a definition. In this case it is helpful to refer to regulation of Art. 16 para 1 of the Transport Law. Pursuant to it a transport contract is concluded by purchasing a travel ticket or by meeting other conditions of access to a means of transport determined by the carrier or by the collective public transport organiser, and if they are not determined—by the fact of taking place in a means of transport. So one should agree that the legislator allows various methods to conclude a transport contract. Not only the purchase of a travel ticket is indicated, but also meeting other conditions of access to a means of transport determined by the carrier or by the collective public transport organiser. It is also necessary to notice that, apart from the presented possibility to determine conditions of access to a means of transport by the carrier or by the collective public transport organiser, provisions of the Transport Law authorise carriers or collective public transport organisers to establish regulations specifying conditions of travellers' service, terms and conditions of check-in, and passenger and luggage transport conditions (Art. 4) as well as authorise municipality councils to determine order regulations referring to the municipal regular transport of persons and to transport of persons and luggage by taxies (Art. 15 para 5) [11].

Pursuant to those regulations related to the collective public transport in the urban transport organised by the Municipality of Białystok § 7 of Order Regulations indicates the meaning of 'appropriate transport document', with respect to services provided by the Białystok urban transport. A transport document confirming the conclusion of a transport contract is 'a valid ticket—consistent with the price list and with the travel tariff zone—purchased, validated, registered when entering the vehicle' or a document entitling to free travelling. Pursuant to § 9 a journey with a ticket encoded on an electronic card to be valid requires the card registration when entering the vehicle. Because of that a ticket encoded on an electronic city card, to be considered an 'appropriate transport document' for the purpose of the Transport Law, must be registered by the passenger, immediately after entering the bus. As a result, with respect to the situation of missing registration, at the entry to the vehicle, of a ticket encoded on an electronic card—i.e. missing appropriate transport document, the authority can determine the amount of an additional charge based on Art. 34a para 2 in connection with Art. 34a para 1 sub-para 1a of the Transport Law [11].

On the basis of the above there are no grounds to state that the introduction of the obligation to register each entry in the situation of purchasing previously and holding a season ticket without a limit of rides violates the collective consumer interest.

3.2 No Violation of Collective Consumer Interest and of Decency

One cannot assume that the introduction of electronic ticket systems and the adoption of the obligation to register a season ticket without a limit of rides at each entering the vehicle violates the collective consumer interest. Pursuant to Art. 24 para 2 of the Act on Competition and Consumer Protection, a practice violating the collective consumer interest is the businessman behaviour, which is illegal or inconsistent with decency, which harms the collective consumer interest [12]. The provision of Art. 24 para 2 of the Act on Competition and Consumer Protection does not protect the economic interest as a consumer need, but his/her legal interest, understood as consumer needs, which were considered by the legislator as worthy protection [5, 13]. The violation of collective consumer interest occurs, if it is found that the legal interest based on appropriate piece of legislation has been violated. This excludes protection based only on the opinion itself that the consumer interest has been violated, if no legal grounds have been indicated, from which such interest should result [14]. And these have not been shown so far. The interest subject to protection does not consist of subjective consumer needs, but the needs, which were provided protection and security by the legislator, which is equated with the legal interest [5, 15].

It is also necessary to draw attention to the fact that it is difficult to show that the obligation of each entry registration is inconsistent with decency. Decency is a general clause, occurring many times in various branches of Polish legislation. As a term, which has not been defined by the law, it became the subject of doctrine interpretation. However, the clause importance is not universal, because it should be decoded in relation to a defined actual state [5]. In the adopted decisions of the Competition and Consumer Protection Court actions aimed at provision of incomplete information, disorientation, causing an erroneous consumer opinion, exploiting his/her ignorance or naivety may be considered inconsistent with decency, so actions which are popularly referred to as unfair, unreliable, deviating from adopted standards of proceeding [5, 16]. Moreover, taking into account the pace of IT technologies development and the origination and application of diverse IT solutions, in many cases related to the user location and identification, it is necessary to consider the updating of standards and adapting them to the changing social–economic environment.

3.3 Request to Use an Electronic Card in the Case of Entitlement to Free Travelling

Another reservation may be related to the introduction of the obligation to get and use an electronic card by persons, who were granted entitlement to travel for free. In the past, in the urban public transport, in which paper tickets were used in the fare collection systems, such persons were travelling based on document issued by various institutions and allowed in the urban public transport as documents entitling to travel—e.g. school or disabled persons IDs, or identity document confirming the age. In the situation of electronic cards implementation it is expected that such card will be issued or received. Like before, the legal grounds for the request to make and use a card as well as to make registrations during journeys by the urban public transport will consist in Art. 16 of the Transport Law. For public transport services users such a requirement has no grounds, but this results most often from unfamiliarity with the fact that free rides for passengers are related at the same time with public financing systems created for this purpose. Because of provision of free rides the carriers receive relevant funds—reimbursement of the lost revenue resulting from honouring the concession/free entitlements and it is necessary to know their amounts—corresponding to the number of transported persons with specific entitlements to travel for free. Therefore a precise registration of the transport volume and of the lost revenue is necessary. Such systems have been used for many years in the public transport of regional and national reach [17], and their application in the urban public transport was limited just by the lack of technical register of the lost revenue. In certain cities the entitlements to travel for free are implemented for the next groups, for example, in recent years free rides were introduced for children and youth, moreover, solutions of free urban public transport are applied on days of high air pollutants content (so-called smog days). Such actions require establishing relevant systems of financing.

The example of railway and regional transport shows that the situation, in which a passenger entitled to travel for free is obliged to register his/her travel (e.g. by collecting a free ticket from the ticket office) is not a new practice in the public transport. This solution has been operating for many years at regional railway carriers, where despite the fact that the passenger holds the entitlement and documents authorising to travel for free, (s)he is obliged to collect each time from the ticket office a single-travel ticket with a 100% concession. The provision of Chapter 3, §17 sub-para 3 of the Koleje Śląskie Transport Tariff may be an example [18], including the clause: '... rides of a child (children) up to 4 years of age are carried out based on a single-travel ticket with a 100% concession together with a document stating the child's age (e.g. child health card)'. This situation, from the point of view of the necessity to perform additional actions before enjoying a free ride, basically is not different from the obligation to register the entry to the vehicle in the urban public transport, and so far it has not raised any objections from the point of view of imposing unfavourable contract conditions. A similar situation exists also in the Warsaw metro, where persons entitled to travel for free, willing to travel by metro, have to collect a free entrance ticket [19].

It can be also added, that the requirement of making additional actions during the use of services, even in a situation that they are provided to a specific social group for free, is nothing new and it was not introduced with the implementation of electronic card systems. Examples of many other services may be given, including such, which are free for the user, where it is necessary to present the required document or to make some action. The use of doctor's services may be given as an example (previous registration is required) or unlimited access cards to recreation facilities, such as swimming pools, gyms, cinemas, where it is always required to register the entry and this way to confirm the previously acquired right to the service.

3.4 Solutions Ensuring the Personal Data Protection

The collection, by a collective public transport organiser, of data on registered check-ins and check-outs in the urban transport vehicles raises concerns related primarily to the privacy protection and personal data protection. The care of personal data security is something natural, especially in the context of more and more universal usage of IT in all areas of life and as a result of personal data gathering and processing on a previously unprecedented scale. It is possible to notice that the concerns for the privacy protection are common to all technologies, which link a device or a card with a specific user and which allow to gather the data on his/her behaviour (mobile telephony, bank systems, i.e. credit and payment cards or cards used in the health service, urban public transport, etc.) [20]. At the same time the issues of personal data protection are very restrictively regulated in the European and Polish legislation. Modern electronic payment systems based on smart cards, such as ŚKUP cards, meet high standards—and in the case where such systems are developed together with banks—safety standards determined by the banking systems. Personal data in electronic card systems are processed in a way guaranteeing meeting requirements set by the Act of 29 August 1997 on Personal Data Protection, together with secondary legislation to that act, which is supervised by appropriate institutions, including the General Inspector of Personal Data Protection [21].

For the system implementing entity it is significant to obtain data about passenger flows for settlements with municipalities and for management of the transport offer on a current basis and for its adaptation to changing needs. Thereby it is necessary to collect and used the data on tickets purchases and registration in individual vehicles. This is the data related to the number and amounts of transactions, and to registered ticket types. Such data may be obtained also with the accuracy of individual cards. To ensure personal data protection in the system of Silesian Public Services Card (ŚKUP) implemented by the Municipal Transport Union of the Upper-Silesian Industrial Region in Katowice, a solution was applied, in which only and exclusively the card number is the identifier (this applies both to personal and bearer cards). In the ŚKUP system, there is no information about such card's user and hence it is not possible to connect the data on the card use with the user personal data, because the personal data is not recorded in the ŚKUP IT system. Also on-board readers in vehicles read the card number and recorded rights to concessionary or free travels, while they do not read from the card and do not register in the system the information about specific card user. Thereby in such cases there is no personal data processing.

Personal data of personalised (personal) cards users is collected in the process of cards ordering and processed for the needs to conclude the contract on the card use and to execute it. It should be clarified, that mBank is the card issuer, to whom Municipal Transport Union entrusted personal data processing for this purpose. While mBank entrusted the personal data processing to the DanubePay company, which is the Clearing Agent of the ŚKUP system and manages the process of cards personalisation. It is just the Clearing Agent's IT system, where the personal data is collected and processed. During the filing of an application for a personalised card, be it on the Customer Portal or in service points, a form is filled, which is automatically transferred via the ŚKUP IT System to the Clearing Agent system, while the data is not saved in the ŚKUP system. Municipal Transport Union, as the ŚKUP system administrator, and other entities participating in the project implementation do not have access to the Clearing Agent system. Thereby it is not possible to connect the data used among others for settlements with the personal data of card users gathered in the Clearing Agent system. This system on the one hand ensures acquisition of relevant data related to the organisation and financial settlement of the public transport, and on the other hand it guarantees personal data protection during their processing.

4 Summary

It is difficult to assume, that the implementation of electronic tickets systems and principles of check-in registration violates the collective consumer interest. The violation of collective consumer interest occurs, if it is found that the legal interest based on appropriate piece of legislation has been violated. This excludes protection based only on the opinion itself that the consumer interest has been violated, if no legal grounds have been indicated, from which such interest should result. And these have not been shown so far. It is also necessary to draw attention to the fact that it is difficult to show inconsistency with decency.

The organiser or carrier are authorised to determine access conditions to a means of transport. This authorisation results from the Transport Law and it is possible to show organisers in Poland, who based on that for many years have been using the rule of obligatory registration of entering the vehicle, even in the situation, where the passenger has previously bought a season ticket without a limit of rides recorded on the electronic card. The application of check-in registration systems may be used to study passenger flows and this in turn to optimise the transport offer on a current basis and to carry out settlements with entities financing the urban public transport from public funds.

The situation, in which big IT projects are financed from public funds should not be assessed very well, when substantial expenditure is incurred and only after such project implementation state authorities perform assessments and raise doubts in the field of applied solutions. It is also possible to expect that in situations raising doubts, but at the same time a specific solution is used in various cities, it would be favourable to develop guidelines or to present best practice, so as to avoid spending substantial funds to achieve defined functionalities, which later on are contested or challenged.

References

1. Website of European Commission. https://ec.europa.eu/transport/facts-fundings/studies_en. For example: Study on Public Transport Smartcards—Final Report AECOM European Commission DG MOVE 1/5/2011
2. Statut Komunikacyjnego Związku Komunalnego GOP, Biuletyn Informacji Publicznej KZK GOP. http://bip.kzkgop.pl/modules.php?name=Content&pa=showpage&pid=2790
3. Umowa z dnia 9 stycznia 2012 r. zawarta pomiędzy KZK GOP a Asseco Poland S.A. i BRE Bank S.A.: Załącznik nr 1.II do Umowy nr DS/1/PZ/1/DO/1/2012 z dnia 9 stycznia 2012 r. – „Opis techniczno-funkcjonalny przedmiotu zamówienia"
4. Decyzja Prezesa Urzędu Ochrony Konkurencji i Konsumentów nr RKT – 2/2018, Katowice 3 grudnia (2018)
5. Klatka, J.: Odwołanie od Decyzji Prezesa Urzędu Ochrony Konkurencji i Konsumentów nr RKT – 2/2018, KZK GOP pismo nr RKT.610.5.2017.MI
6. Dydkowski, G., Tomanek, R., Urbanek, A.: Taryfy i systemy poboru opłat w miejskim transporcie zbiorowym. Wydawnictwo Uniwersytetu Ekonomicznego w Katowicach, Katowice (2018)

7. Dydkowski, G.: Integracja transportu miejskiego. Prace Naukowe Akademii Ekonomicznej im. K. Adamieckiego w Katowicach, Wydawnictwo Akademii Ekonomicznej w Katowicach, Katowice (2009)
8. Wyszomirski, O.: Modern urban transport systems in Europe. In: Burnewicz, J. (ed.) Innovative Perspective of Transport and Logistics, pp. 189–200. Wydawnictwo Uniwersytetu Gdańskiego, Gdańsk (2009)
9. Szpil, M., Rumas, M.: Badanie opinii pasażerów na temat nowej taryfy KZK GOP i systemu ŚKUP. TNS Polska Oddział Katowice dla KZK GOP, TNS Polska S.A., Katowice (2012)
10. Ustawa z dnia 15 listopada 1984 roku Prawo przewozowe, tekst jednolity Dziennik Ustaw z 2017 r. poz (1983)
11. Uzasadnienie Wojewódzkiego Sądu Administracyjnego w Białymstoku, sygnatura akt I SA/ Bk 82/14. http://www.orzeczenia-nsa.pl/szukaj?q=&_symbol=639_6391&show=symbol
12. Ustawa z dnia 16 lutego 2007 roku o ochronie konkurencji i konsumentów Dziennik Ustaw 2007 Nr 50 poz. 331 z póź. zmianami
13. Uchwała Sądu Najwyższego z dnia 13 lipca 2006 r(, sygn.. Akt III SZP 3/06.
14. Kohutek, K., Sieradzka, M.: Ustawa o ochronie konkurencji i konsumentów. Commentary. Wolters Kluwer Polska Sp. z o.o, Warszawa (2008)
15. Banasiński, C., Piontek, E. (red/eds.): Ustawa o ochronie konkurencji i konsumentów. Commentary. Lexis Nexis Polska Sp. z o.o., Warszawa (2009)
16. Website of the Court Verdicts Analysis System. https://www.saos.org.pl/. Judgment of SOKiK of 3 April 2013, file No XVII AmC 6071/12.
17. Ustawa z dnia 20 czerwca 1992 roku o uprawnieniach do ulgowych przejazdów środkami publicznego transportu zbiorowego, tekst jednolity z 2018 roku, poz. 295
18. Taryfa przewozowa Koleje Śląskie sp. z o.o. http://kolejeslaskie.com/wp-content/uploads/2018/09/Taryfa-Przewozowa-TP-K%C5%9A.pdf
19. Website of Public Transport Authority. http://www.ztm.waw.pl/pytania.php?c=112&l=1
20. Urbanek, A.: Automated fare collection systems based on check-in and check-out—premises of implementation in urban public transport. Archiv. Transp. Syst. Telemat. **10**(3), (2017)
21. Ustawa z dnia 29 sierpnia 1997 r. o ochronie danych osobowych, Dz. U. 1997 Nr 133 poz. 883 wraz z późniejszymi zmianami

Can the Metropolitan Rail System Hamper the Development of Individual Transport? (Case Study on the Example of the Szczecin Metropolitan Railway, Szczecin, Poland)

Krystian Pietrzak and Oliwia Pietrzak

Abstract Rail transport is becoming an increasingly common mode of transport supporting agglomeration/metropolitan traffic. It is affected by a variety of factors, such as substantial degree of punctuality and safety, possibility of transporting a significant number of passengers at the same time, limited negative impact on the natural environment and the quality of being independent from road congestion. The emerging metropolitan rail networks can effectively counteract the uncontrolled development of individual transport, thus promoting the use of public transport in handling passenger traffic in urban and metropolitan areas.

The primary purpose of the chapter was to identify current barriers hampering the use of public transportation by metropolitan residents and show the possibility of reducing those barriers by a metropolitan railway system on the example of the Szczecin Metropolitan Railway investment project in Poland.

To this end, the following research methods were applied: critical evaluation of the relevant literature, documentary method, survey method, and individual in-depth interviews. In addition, the results of research carried out by the Authors during the implementation of the "Concept of the Szczecin Metropolitan Railway" (2012) and "Feasibility Study of the Szczecin Metropolitan Railway" (2014–2015) were used.

The area of research was the Szczecin Metropolitan Area (SMA), Poland.

Keywords Rail transport · Passenger transport · Public transport · Sustainable development · Urban railway · Metropolitan railway · Szczecin Metropolitan Railway

K. Pietrzak (✉) · O. Pietrzak
Maritime University of Szczecin, Szczecin, Poland
e-mail: k.pietrzak@am.szczecin.pl; o.pietrzak@am.szczecin.pl

M. Suchanek (ed.), *Transport Development Challenges in the 21st Century*, Springer Proceedings in Business and Economics,
https://doi.org/10.1007/978-3-030-50010-8_16

1 Introduction

Transport is an important part of human life. Due to demographic, economic, and lifestyle shifts people are traveling more often and further, and the time they spend inside means of transport is becoming longer. Therefore, mobility is proving to be one of the most important human needs of the twenty-first century [1]. Passenger transport and goods traffic also constitute an important basis for social participation and the prosperity of the population [2]. Citizens who are unable to access public transport are excluded from participating in economic and social activities [3]. Therefore, modern transport can be considered as:

- generator of economic development,
- important factor for improving the well-being of people,
- tool to ensure inter-territorial ties,
- indispensable tool for trade exchange in the era of globalization.

At the same time, transport has become very harmful to the environment. Its development has exerted negative impact on the environment through, among others, noise, vibration, accidents, area needs, congestion, and energy intensity [4]. Currently, cities and agglomerations are facing numerous problems caused by the increasing level of vehicle-to-population ratio [5]. The dynamic development of urbanized areas and, in particular, the uncontrolled spatial development, further deepen this problem.

Social expectations for improving the conditions of life in cities and current increase in mobility pose a difficult challenge for organizers of urban transport [6]. In general, cities and agglomerations are trying to make automobile travel less attractive and increase the use of public transport [7]. The organization and development of public transport in metropolitan areas is an important task of self-governments, because the efficacy of transport systems in highly urbanized areas translates directly into the rate of their economic development [8].

The primary purpose of the chapter was to identify current barriers hampering the use of public transport by metropolitan residents and show the possibility of reducing those barriers by a metropolitan railway system on the example of the Szczecin Metropolitan Railway investment project (SMR) in Poland.

An important part of the research was to answer the following questions:

- What is the current state of passenger transport system in SMA?
- What solutions will the SMR investment project offer to its future passengers (SMA residents and visitors)?
- What will be the target passenger system in the SMA (after starting the SMR)?

The chapter presents the results of the research, which was carried out in two stages:

- first, surveys were conducted among SMA residents (2013, 2015, 2017),
- second, in-depth individual interviews were conducted with a deliberately selected group of SMA residents (2018).

2 The Role of Railways in Serving Passenger Traffic

Under favorable conditions, the numerous characteristics of rail transport may allow it to compete with other modes of transport. They include [9, 10]:

- Travel safety,
- High speed,
- High degree of punctuality,
- Limited impact on the natural environment,
- Low degree of vulnerability to weather conditions,
- Relatively low degree of occupancy of the area,
- Independence from traffic congestion, and.
- Ability to transport a significant number of passengers at once.

The aforementioned characteristics of railway transport cause it to be an increasingly common element of city passenger transport systems. The railway may affect the increase in the competitiveness of public transport in relation to individual transport. Furthermore, urban development harmonized with urban railway systems with a higher service frequency should be a key factor in the strategic implementation of low-carbon urban public transportation corridors [11]. The delayed timing of urban railway introduction could contribute to a negative spiral of motorization, which would provoke a severe increase in CO_2 emissions from the transportation sector [11].

There are notions characterizing the functioning of urban railway in cities and metropolises: Light Rail Transit (LRT) and Light Rail System (LRS). LRT/LRS are understood as a solution which makes use of the railway to integrate the passenger transport system [12] into urban or metropolitan areas. This system provides high-quality, high-speed, and environmentally friendly public transit service on established trunk corridors linking major trip generators, regional centers, and county cores [13]. The SMR may become the core of the transport system in the region, providing connections between main generators of passenger traffic.

The LRT/LRS system is an element of railway services, whose role is to support the development of urbanized areas and reduce traffic congestion [14]. The implementation of LRT/LRS is an essential factor encouraging the shift from individual to public transport [15]. Therefore, it may be pointed out that rail transport, including the LRT/LRS system, may be a significant and useful tool to create sustainable and friendly cities and agglomerations.

To recap, LRT/LRS systems offer an advantage over traditional means of transport used in urban and agglomerations traffic, because they may help to create main transport corridors [16], which provide connections between important transport nodes of urbanized areas.

When assessing the importance of particular modes of transport for the creation of a sustainable passenger transport system in cities and metropolises, one can point to the importance of rail transport. Due to the low susceptibility to weather conditions as well as separate infrastructure and independence from the road congestion,

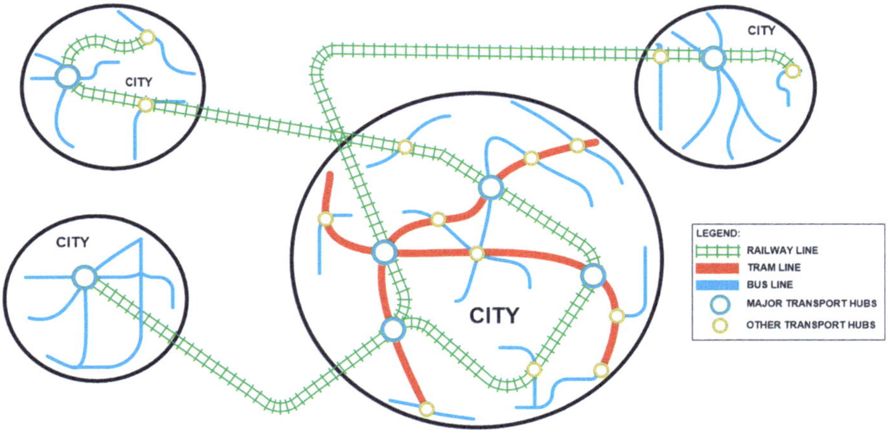

Fig. 1 The role of particular modes of public transport in the metropolitan transport system (Own source)

railway may maintain high frequency and punctuality, performing regardless of the time of day (also during communication peaks). It should be remembered, however, that there are some limitations—the train will not be able to fully replace the functionality of a car or bus. For instance, it is unable to help passengers commute directly from home to work. The role of rail transport should therefore be directed primarily toward servicing large passenger flows between main transport hubs in the region, including the connections between neighboring cities and the main urban center of a given metropolis.

Designing a sustainable transport system for a city and a metropolis usually requires the involvement of other modes of transport to support the functions of the railway. Where possible, other forms of public transport, such as a tram, bus, metro, should be preferred. Figure 1 shows the diagram of the functioning of the system in question.

The use of rail transport in urban and metropolitan transport systems, in addition to activities oriented at changing the transport behavior of its users and improving the quality of the environment, may also bring other benefits. The presence of a modern and efficient railway system may affect the competitiveness and prestige of a given area.

3 Szczecin Metropolitan Area (SMA) as a Research Area

The development of metropolitan areas is one of the most important factors which contribute to the competitiveness of the region in the domestic and international market. The Szczecin Metropolitan Area (SMA) is located in the West Pomeranian Voivodeship and includes 13 cities and municipalities. Key SMA data are as follows:

- Area—2795 km^2 (representing 12.21% of the area of the West Pomeranian Voivodeship).
- Population—686,582 people (representing 40.4% of the total population of the West Pomeranian Voivodeship).
- Economic potential—45.8% of regional economic entities are located within the SMA.

The primary objective of the SMA is to strengthen the cooperation between local self-governments (in particular, between Szczecin and neighboring units) and to develop and implement a coherent urbanization system, integrated transport system, communication network and to strengthen its social capital.

The SMA passenger transport system consists of several different subsystems:

- Transport system of the main city of the SMA (Szczecin),
- Transport systems of medium-sized cities (Stargard, Goleniów, Gryfino, Świnoujście, Police),
- System of connections between the indicated urban centers as well as these centers and small towns and rural areas.

Problems which the SMA transport system is facing are similar to those occurring in other urban and metropolitan centers in Poland, Europe, or the world. The most important ones include too many individual vehicles, unsustainable modal split, traffic congestion, or air pollution.

The current division of tasks between particular modes of transport shows that road transport (including individual transport) plays a dominant role. As there is no proper infrastructure or organization of the system, the railway transport only plays a marginal role in servicing passenger flows in the SMA area (Fig. 2).

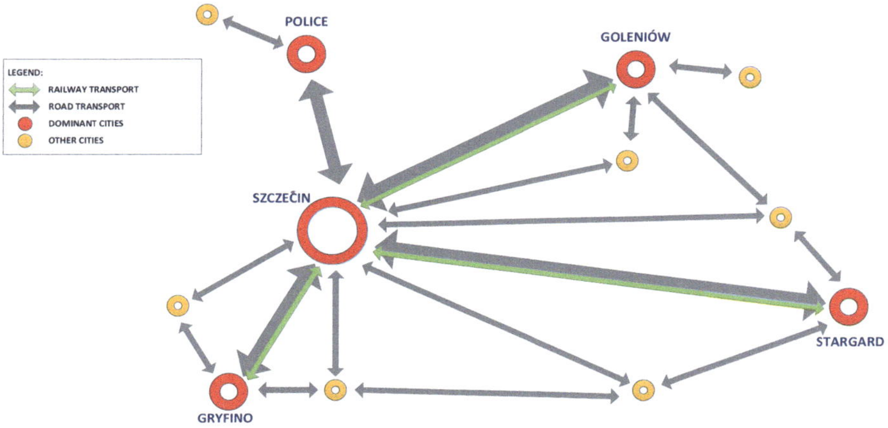

Fig. 2 Diagram of passenger transport system in SMA—current state (Own source)

4 Szczecin Metropolitan Railway (SMR): Objectives and Solutions

The main objective of the SMR is to create an efficient transport system implementing passenger connections in the Szczecin Metropolitan Area [17]. An important solution is to modernize the existing railway network within the SMA area and to create the main public transport axis thereupon. This network will connect the capital of the voivodeship, Szczecin, with other cities in the area: Stargard (railway line 351), Goleniów (line 401), Gryfino (line 273), and Police (line 406). The SMR will primarily be responsible for handling passenger flows between the main transport hubs in the region. The assumptions adopted for the purposes of the feasibility study of the investment project (2014/2015 and update 2016/2017) [17] show that the SMR will be fully integrated with other modes of transport, both public and individual. The main solutions of the SMR investment project are as follows:

- Construction of interchanges connected with other modes of transport,
- Construction of new Park&Ride parking lots (over 1500 new parking spaces),
- Construction of new Bike&Ride parking lots (over 1280 new parking spaces),
- Modification of the course of bus and tram lines and adjusting them to the SMR,
- Coordinating the timetables of bus and tram lines and adjusting them to the SMR,
- Integrated ticket system for all types of public transport in the SMA,
- Purchase of modern trains,
- Adaptation of all train stops to the needs of families with children and the disabled.

The SMR will provide its users with high degree of punctuality, favorable travel time and high frequency of running vehicles adapted to their needs in terms of mobility. Its purpose is to introduce changes, so that public transport will be able to compete with individual transport, consequently reducing the number of journeys made in private cars. The SMR provides for passenger services at 40 diverse stops: international (1), agglomeration (5), urban (23), or local (11). Some will include completely new (9), revitalized (after being out of use for a long time) and modernized stops.

It is the opinion of the authors that another important issue is the additional effect which the construction of the SMR will exert—there will be as many as 26 SMR stops in the area of Szczecin (currently, there are only 9), which will also allow it to function as an urban railway. With such organization, the rail system may significantly strengthen the role of public transport in servicing typically urban passenger flows.

It is worth pointing out that the SMR, implemented in accordance with the economic and social analysis carried out as a part of the feasibility study, successfully applied for financial support offered by EU funds. The indicators of economic and social efficiency were higher than the limit values. In the case of ENPV, it was PLN 751.8 m, and IRR—11.9%. In 2018, the SMA received EU grants to help finance

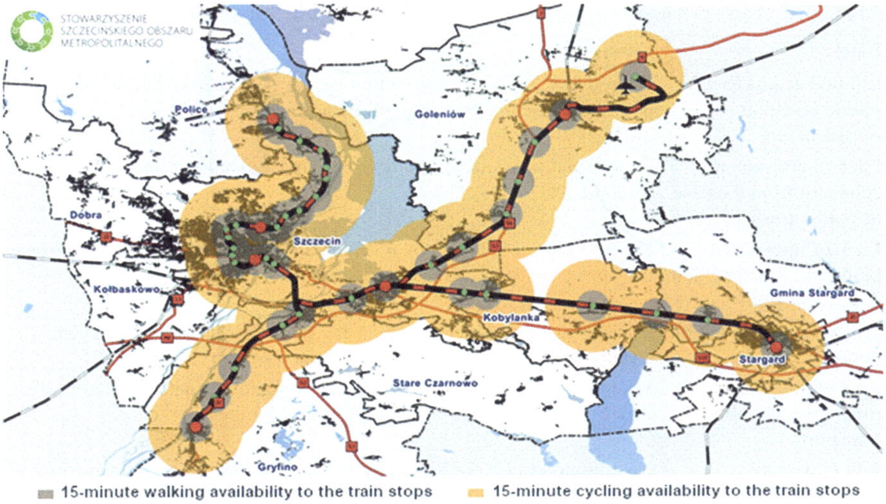

Fig. 3 Walking and cycling availability of SMR stops [17]

the project. The same year, the SMR investment process began. We may assume that the system will be fully operational in 2022. Figure 3 presents the expected availability of SMR stops (for walking and cycling).

5 Szczecin Metropolitan Railway: Possibilities of Reducing the Development of Individual Transport

Several studies were conducted in order to identify current barriers hampering the use of public transportation by metropolitan residents and show how to reduce these barriers for the metropolitan railway system on the example of the investment project of the Szczecin Metropolitan Railway.

First, in 2013, 2015, and 2017, the authors conducted a survey among the residents of the SMA and the West Pomeranian Voivodeship. The study involved a group of respondents: 1752 people in 2013, 1528 in 2015, and 1887 in 2017. The wide scope of the research concentrated on the passenger transport system in the region, both in terms of individual and public transport. One of the objectives of the research was to identify barriers hampering the use of public transport in the area of SMA and the West Pomeranian Voivodeship. The key barriers pointed out by respondents are shown in Table 1. More results of the surveys are discussed in [18, 19].

Respondents also pointed to other, often incidental factors, such as illness, sudden event and unwellness. Due to the marginal nature of these answers, the authors left them out of further analyses.

Table 1 Key barriers hampering the use of public transport by users of the SMA transport system

Barriers	Description
Limited access to public transport infrastructure	No nodal or linear infrastructure (e.g., stop or railway line) near the point of beginning or ending of the journey
Lack of convenient connections in the case of public transport	Lack of public transport connections adjusted to the needs of respondents
Lack of integration between individual and public transport	No possibility of leaving own vehicle in a designated area and continue the journey by public transport
Too long travel time by public transport	Travel time is not satisfactory for the passenger, thus the passenger usually decides to use their own vehicle
Lack of integration of timetables in public transport	Waiting time to change from one public vehicle to another is too long, which increases the duration of the entire journey
Inflexible/not integrated ticket tariff	No common ticket for various means of transport (e.g., railway, tram, bus), lack of integrated tariff between public transport and complementary solutions (e.g., P&R)
Lack of adaptation to the needs of families with children	Lack of facilities in vehicles (e.g., suitable space for baby strollers, low-floor vehicles), lack of facilities at stops (e.g., low curbs, elevators) or unattractive ticket tariff (e.g., discounts for journeys with children, family tickets)
Lack of adaptation to the needs of the disabled	Lack of facilities in vehicles (e.g., suitable space for wheelchairs, low floor vehicles), lack of facilities at stops (e.g., low curbs, elevators)
Convenience of using individual transport	Individual feeling of the passenger toward his own vehicle; it may refer to objective features, e.g., door-to-door travel, adjustment of departure time to individual needs, as well as subjective assessment of vehicle features, e.g., dual zone air conditioning, comfortable seats, tinted windows
Limited promotion of public transport	No advertising campaign, no temporary promotions
Habits, routine	Passengers often choose well-known solutions without looking for new travel opportunities
Too expensive tickets	Ticket prices are uncompetitive compared to individual transport

Further, the solutions adopted in the SMR investment project were indicated. They are the result of the research carried out as a part of developing the SMR feasibility study (as indicated in Part 4 of the chapter, the research was carried out in 2014–2015 and updated in 2016–2017).

The final stage of the research consisted in individual in-depth interviews in 2018 with 30 respondents. They were selected from among people living in the SMA area, who did not use public transport there or only used it sporadically.

The respondents received a list of barriers constituting the result of surveys and a list with solutions adopted for the SMR investment project. Their task was to:

- Select and indicate barriers that concern them, and.

- Indicate which of the adopted SMR investment solutions can eliminate individual barriers hampering the use of public transport.

As a result, a matrix of SMR impact on the barriers to the use of public transport in the SMA area was obtained (Fig. 4), in which:

- Barriers hampering the use of public transport (survey results) are shown in red.
- SMR investment solutions (from the SMR feasibility study) are shown in orange.
- Sustainable transport policy goals (local, domestic and European) are shown in green.

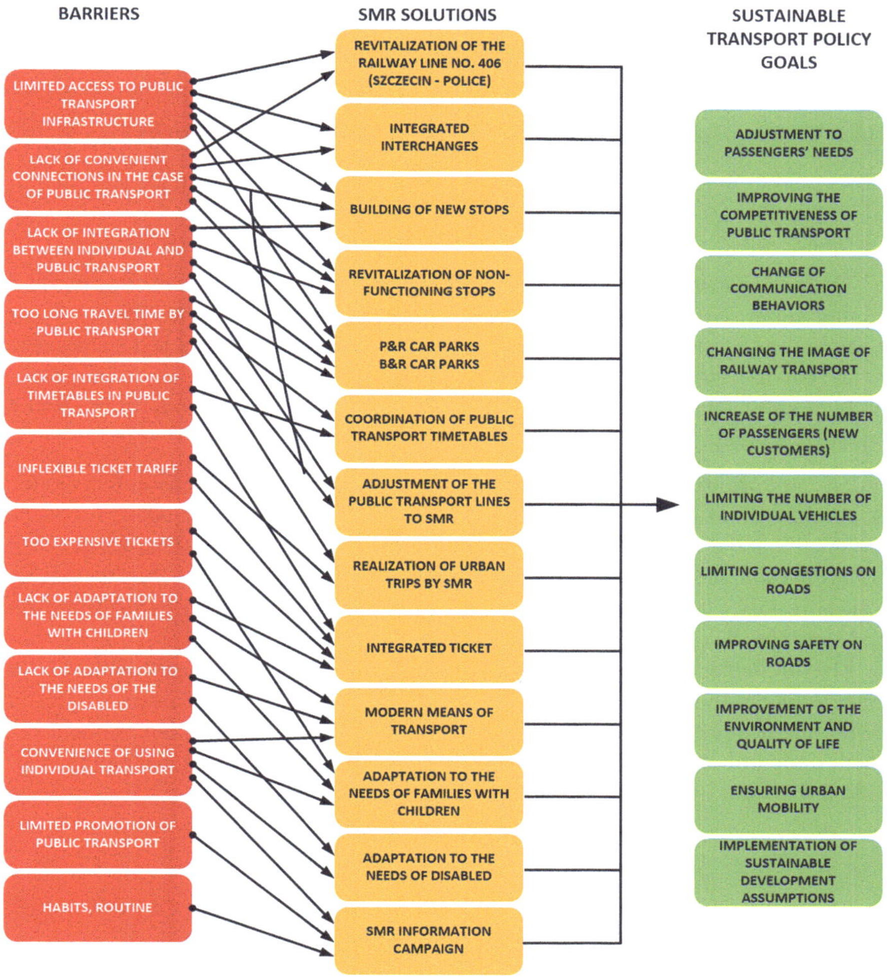

Fig. 4 Diagram of SMR impact on barriers to the use of public transport within the SMA (Own source)

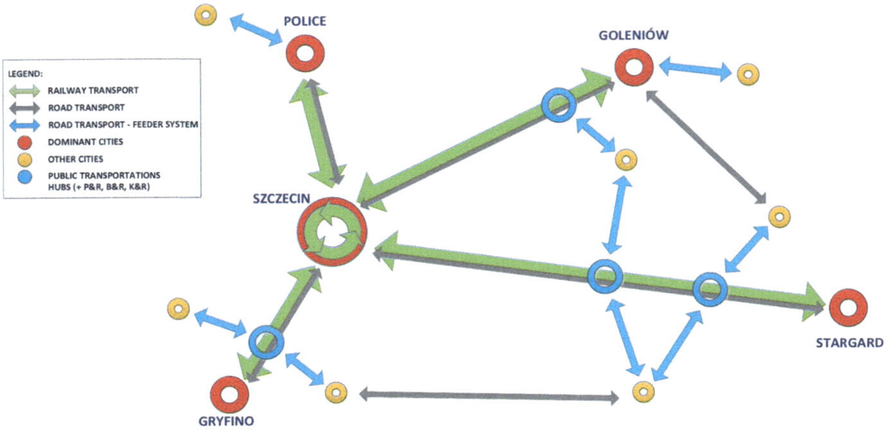

Fig. 5 Diagram of passenger transport system in SMA—target state (Own source)

As a result of research carried out in the SMA area, it was shown that:

- There are barriers hampering the use of public transport; they are diverse because they relate to infrastructure, economic, organizational, and social issues,
- Solutions adopted as a part of the SMR construction project may be an answer to the barriers indicated by residents,
- Individual solutions adopted in the SMR investment project may limit several barriers hampering the use of public transport.

The authors are of the opinion that the developed matrix may constitute a model, as the impact of SMR solutions on other barriers cannot be excluded. It can be assumed that the impact will have a wider reach. The implementation of individual SMR investment solutions may help to achieve specific goals of the sustainable transport policy, and consequently increase the competitiveness of public transport and limit the development of individual transport in the SMA area.

The implementation of the SMR investment solutions may contribute to the change of the modal split in the SMA area. For the sustainable development of SMA, the desired system is the one where public transport plays a dominant role (with a significant share of rail transport—SMR). Road transport, including individual transport, serves only as a supporting system (Fig. 5).

6 Conclusion

The SMR as a solution combining significant transport capacity with limited susceptibility to transport congestion may decide on the success of public transport over individual transport. It may be observed in the following relations:

- City–city,
- City–suburban areas,
- To a limited extent, intracity relations (e.g., service by rail connections within the capital of the region—Szczecin).

This study shows that the SMR can bring benefits for the residents of the entire area included in the SMA as well as the residents of Szczecin. The former may be able to enjoy rail connections (competitive to individual transport) between the cities of the metropolitan area and the cities through which the SMR network will run. When traveling to Szczecin, they will be able to choose one of many stops near the destination of their journey. The SMR running through the area of Szczecin, along with as many as 26 stops on its territory, may gain additional functionality—urban railway. Residents of the city can use the SMR when moving between its districts. The achievement of the target modal split depends, however, on the implementation of a full set of solutions that limit or eliminate barriers to the development of public transport in the SMA identified by the authors. If the implementation of certain SMR solutions is incomplete, they may result in delays or failures to achieve the assumed objectives. Consequently, limiting the development of individual transport in the SMA area using the metropolitan railways may be slower or insufficient.

References

1. Skrucany, T., Kendra, M., Skorupa, M., Grencik, J., Figlus, T.: Comparison of chosen environmental aspects in individual road transport and railway passenger transport. Proc. Eng. **192**, 806–811 (2017)
2. Saighani, A., Sommer, C.: Method for an economical assessment of urban transport systems. Transp. Res. Proc. **37**, 282–289 (2019)
3. Adhvaryu, B., Patel, M.: Is public transport in Ahmedabad inclusive? Econ. Polit. Wkly. **54**(8), 17–20 (2019)
4. Dolinayova, A.: Factors and determinants of modal split in passenger transport. Horizon. Rail. Transp. Sci. Pap. **2**(1), 33–39 (2011)
5. Hovavko, I.Y.: Economic analysis of traffic jams in Moscow. Public administration. Electron. Gazette. **43**, 121–134 (2014)
6. Pomykala, A.: Effectiveness of urban transport modes. MATEC web of conferences, vol. 180, 03003. (2018)
7. Varghese, V., Adhvaryu, B.: Measuring overcrowding in Ahmedabad buses: costs and policy implications. Transp. Res. Proc. **17**, 145–154 (2016)
8. Ważna, A., Ważny, K.: Organisation of public transport in metropolitan areas. Res. J. Univ. Gdańsk. Transp. Econ. Log. **67**, 53–63 (2017)
9. Pietrzak, O., Pietrzak, K.: The role of railway in handling transport services of cities and agglomerations. Transp. Res. Proc. **39**, 405–416 (2019)
10. Pietrzak, K.: Analysis of the possibilities of using "Light Freight Railway" for the freight transport implementation in agglomeration areas (example of West Pomerania Province). Transp. Res. Proc. **16**, 464–472 (2016)
11. Tsumura, Y., Nakagawa, D., Matsunaka, R., Oba, T.: A study on the timing of Asian metropolitan railway development with motorization and its effect on CO2 emissions from transportation. IATSS Res. (2019). https://doi.org/10.1016/j.iatssr.2019.01.001

12. Huang, J., Xu, Y., Wang, C., Xiong, L., Hai, J.: Research on problem of urban regeneration and optimization of land-use around light rail transit in Nanhai District of Foshan City. Transp. Res. Proc. **25**, 3055–3065 (2017)
13. https://nacto.org/docs/usdg/lrtserviceguidelines_vta.pdf
14. Zhou, L., Liu, Y.-J., Yu, L., Liu, Y.: Research on the spatial-system-based rail transit. Systems of the world cities. Proc. Eng. **137**, 699–708 (2016)
15. Ortega, J., János, T., Tamás, P.: Estimation of parking needs at light rail transit system stations. Conference on Transport Sciences 2019, at Győr-Hungary. ISBN 978-963-8121-85-1 (2019)
16. Huang, R., Moudon, A.V., Zhou, C., Stewart, O.T., Saelens, B.E.: Light rail leads to more walking around station areas. J. Transp. Health. **6**, 201–208 (2017)
17. http://www.skm.szczecin.pl
18. Pietrzak, O.: Systemy transportu pasażerskiego w regionach—funkcjonowanie, kształtowanie, rozwój. Przykład województwa zachodniopomorskiego. Wydawnictwo Bel Studio, Szczecin (2015)
19. Pietrzak, O.: System transportu pasażerskiego województwa zachodniopomorskiego w ocenie jego użytkowników. Prob. Transp. Log. **35**(3), 131–141 (2015)

Generalized Cost of Daily Trips on the Example of Public Transport and Private Car Selection

Zuzanna Kłos-Adamkiewicz

Abstract Transport users frequently face the dilemma of maximizing the utility of usage of two resources—time and money in relation to the transport services. The perceived value of the transport services for passengers is the result of what they have received in relation to the costs of transport service. It expresses the concept of generalized cost, which includes financial costs, time costs, discomfort, and other elements that can be perceived in general by the user as costs.

The chapter presents the generalized cost on the example of selection of private car and public transport. Car users, while comparing the ability to move by car or by public transport, often do not take into account the full cost of car ownership and usage. While in urban transport homogeneous rates for different tariffs are adopted, whereas in the case of individual transport costs are not limited to the fuel price. The concept of generalized cost is presented in the chapter, based on market research carried out in Szczecin and the proposed method of calculating these costs in transport.

Keywords Generalized travel cost · Urban transport · Passenger car · Cost analysis

1 Introduction

Every journey is a process of people moving in a certain way between designated travel destinations (starting and ending points) that can be divided into particular stages, which are related to the choice of the travel mode (e.g. on foot, using bicycle or using a specific mode of transport). Therefore, each journey can be treated as a

Z. Kłos-Adamkiewicz (✉)
Faculty of Management and Economics of Services, University of Szczecin, Szczecin, Poland
e-mail: zuzanna.klos-adamkiewicz@usz.edu.pl

M. Suchanek (ed.), *Transport Development Challenges in the 21st Century*, Springer Proceedings in Business and Economics,
https://doi.org/10.1007/978-3-030-50010-8_17

chain of displacements implemented as part of the transport of people, while each displacement is identified as a specific stage.

The chapter presents the methodology of assessing generalized costs and answers the question about the differences between the type of costs that appear while using private car and public transport. It introduces to the subject concerning estimation of full costs of transportation. The idea of generalized costs appears in transport literature, though very often it's not presented in details. It is important to underline, that presented case shows how to estimate those costs in practice, using specific data.

Research in the area of generalized travel costs is important in the case of assessing the attractiveness of individual transport modes, in this case individual transport and public transport. Transport users take into account different variants of the service in the selection of means of transport, taking into consideration elements such as price, quality and time of travel. The usability theory assumes that they are able to clearly formulate their preferences and evaluate alternatives, and the consequences of each solution might be estimated [1].

2 Generalized Cost of Journey: Theoretical Approach

In recent years, there has been an increase in the significance of the notion of the generalized cost, both in theory and in practice. There is no clear definition of this concept in literature, but very often in economic theory it is used in numerous studies and researches. The generalized cost is one of the categories of transport economics, and in particular the analysis of prices and costs.

One of the authors of the definition of the generalized cost of travel is Kenneth Button, who states that this is a measure of measurable character (often presented in the form of a line chart), presenting the total travel costs [2].

K. Button presents a simplified pattern of generalized costs based on the formula (Eq. 1) [3]:

$$G = g\left(C_1, C_2, C_3, \ldots C_n\right) \tag{1}$$

where: G—generalized cost, C_1, C_2, C_3, ... C_n—different travel costs (money, time, other).

The generalized cost of travel is related to the time devoted to travelling (often perceived as a loss of time) as well as the costs associated to the movement.

Another example of the perception of generalized costs was presented by Trujillo and Nombella [4]. They paid particular attention to the value of time that should be spent on the travel. According to their formula, the generalized cost is (Eq. 2):

$$G = p + \left(t \times V\right) \tag{2}$$

where: G—generalized cost, p—price of transport service, t—time, V—value of time.

U. Marchese presents the generalized cost as the sum of costs/prices in transport and the value of time spent on travelling. According to him, costs of transport and a measurable value of time are homogeneous and they are elements that can be added to each other. It follows that it is the sum of monetary and nonmonetary travel expenses. Monetary costs include: fuel, work, insurance, and maintenance. Nonmonetary costs include elements that are the sum of qualitative attributes resulting from the travel, which do not have a direct monetary indicator, but play an extremely important role in the choice of mode of transport and how it's being assessed and perceived. It can be included, among others: reliability, speed, security, punctuality, etc. According to this author, the model of generalized cost is as follows (Eq. 3) [5]:

$$G = c + u\left(m_1, m_2, m_3, \ldots m_n\right) \tag{3}$$

where: G—generalized cost, c—monetary costs, u (m_1, m_2, m_3, ... m_n) —nonmonetary costs, being a function of the attributes of transport mode.

The generalized cost includes all possible costs that the user has to cover in order to overcome the path from the point of start to the ending point of his journey. It includes, depending on the chosen means of transport: operating costs, travel time costs taking into account the level of nuisance of individual travel stages, which may be expressed in financial, time or other form corresponding to the value of time.

3 Valuing Time

As noted by many authors referring to the concept of generalized travel costs (Button, Trujillo, Gombella), the important element that determines the size of this cost is the time necessary to complete the journey. The literature distinguishes different definitions of time that relate to travel (Table 1).

Different ways of perceiving time can be noticed (subjective or objective character) and how this information can be used for economic analysis. The most general concept is the time of the clock, which has an objective character and enables real time estimation.

In the case of travel time, which is a very broad concept, one can distinguish time perceived by the user (subjective in nature), time devoted to travelling related to paid work (which affects our income) or personal travel time related to everyday mobility. The time that is devoted to performing professional duties should be measured on the basis of real time, while the personal travel time is subjectively assessed, in most cases differently from the time (objective) time. The generalized cost included in the table contains monetary elements related to the cost of travel and nonmonetary related to travel time, which translates into their measurable,

Table 1 Defining time

Time concept	Definition	Implication
Time of the clock	Travel time in objective terms, determined by the clock	The way in which time is measured
Travel time, including	Time spent on the journey	The most widely perceived time category and the least detailed aspect of its perception
Perceived (cognitive) time	Journey time experienced by the traveller, which may significantly differ from the clock time (time in a subjective approach)	It reflects the comfort of travelling
Time spent on travelling related to work	Time spent by the employee to perform their basic professional duties, travel time included in working time (deliveries, trips to work-related places)	This type of travel often has a high hourly value
Personal travel time	Time allocated for daily trips (daily travel to work, school, etc.)	This type of time is given the most attention in economic analysis
Generalized cost	A combination of travel time and financial costs associated with the travel	This is an example of including travel time in traffic models (e.g. cities transport models indicating travel time at different times of the day)

Source: own elaboration based on [6]

cost-intensive nature. The considerations regarding the value of time are different depending on the chosen means of transport and different from the needs of individual users.

4 The Method of Measuring the Generalized Cost for Travel Made by Public and Individual Transport

Taking into consideration of the specific features of the functioning of public transport, which involves many features that must be met to meet the needs of its users, it affects the way of measuring the generalized cost. In the process of identifying nonmonetary elements of this cost, the travel behaviour of residents should be included, their preferences or the perceived value of travel time. The same applies to the use of passenger cars for daily trips. Each of the presented ways of moving is diverse in terms of the components of the generalized cost. This cost, in addition to its monetary components (travel cost) includes the subjectively estimated real time of the user's journey (it can be referred to the perceived time, time related to paid work and personal travel time).

The starting point for determining the methods of measuring the generalized cost for journeys made by means of public transport and car should be the separation of travel stages (displacement chains) (Fig. 1).

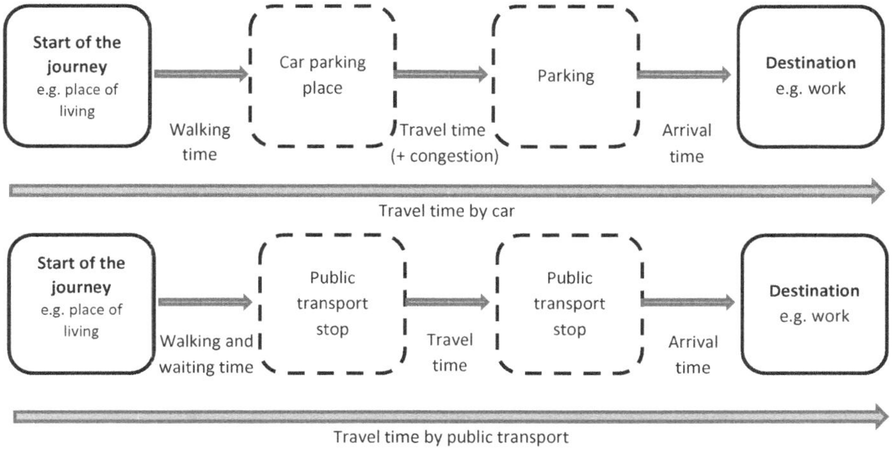

Fig. 1 The travel process with usage of individual and public transport (Source: own elaboration)

In relation to the components of the generalized cost of travel by car and public transport, the following monetary and nonmonetary costs of this type of travel can be distinguished (Table 2).

In the case of stages of car travel, elements affecting the overall travel time that should be distinguished:

- time to get to the parking place (or garage) of the car;
- driving time, which should also take into account the time of stopping or lengthening the journey resulting from congestion;
- time to look for a parking space;
- time to reach the destination of your trip (which may vary depending on the choice of parking space).

Considering the previously presented formulas at a generalized cost and accepting the previously described monetary costs of travel by car and other nonmonetary costs, the following formula for the generalized cost of travel by car can be presented (Eq. 4):

$$G_{\mathrm{so}} = K_{e} + K_{\mathrm{park}} + K_{\mathrm{npos}} \tag{4}$$

where: G_{so}—generalized cost of travelling by car [PLN], K_{e}—car maintenance costs [PLN], K_{park}—parking costs [PLN], K_{npos}—nonmonetary costs [PLN].

Formulas for individual components of the generalized cost of travel by car (Eqs. 5 and 6):

$$K_{\mathrm{e}} = K_{\mathrm{p}} + K_{\mathrm{n}} + K_{\mathrm{u}} + K_{\mathrm{pt}} + K_{\mathrm{k}} + K_{\mathrm{pz}} \tag{5}$$

Table 2 Components of the generalized travel cost

	Monetary costs	Nonmonetary costs
Passenger car	• Car operation – costs of consumed fuel; – costs of repairs resulting from the use of the car and the wear of its components (which are diversified in terms of driving style, type of car, age, etc.); – insurance costs; – costs of annual reviews; – credit, loan or leasing costs; – other costs related to car; • parking costs (Parking costs may include both parking at the destination (e.g. in the paid parking zone in the city) as well as at the source of the travel, in the case of renting a parking space.)	• These are all of the elements of travel time resulting from the implementation of its subsequent stages (walking time to reach the car, driving time, time to look for a parking space, walking time to destination point)
Public transport	• Costs of travelling by public transport (single tickets or season tickets)	• These are all of the elements of travel time resulting from the implementation of its subsequent stages (time to reach the bus stop, waiting time at the bus stop, driving time, time needed for changing the vehicle, arrival time)

Source: own elaboration

where: K_p—fuel costs [PLN], K_n—costs of repairs [PLN], K_u—insurance costs [PLN], K_{pt}—review costs (technical etc.) [PLN], K_k—loan/leasing costs [PLN], K_{pz}—other costs [PLN].

$$K_{\mathrm{npos}} = \left(t_{\mathrm{do}} + t_{\mathrm{j}} + t_{\mathrm{park}} + t_{\mathrm{doc}}\right) \times k \tag{6}$$

where: t_{do}—travel time from starting point to park place [min], t_j—journey time (including congestion) [min], t_{park}—time needed to find park space [min], t_{doc}—travel time to destination point [min], k—unit cost of travel time [PLN/h].[1]

The amount of operating costs, parking costs and non-monetary costs should be calculated on a full-year basis.

Within the components of the time that should be spent on travelling by public transport, it can be distinguished:

- time to get to the public transport stop;
- waiting time for the public transport vehicle;
- driving time, which should also take into account the time of stopping or lengthening the journey resulting from congestion;

[1] In the analysis the value of unit cost of travel time for work travels in 2018, which is 34.38 PLN, was based on [7].

- time needed for possible change of means of transport (sometimes travellers are able to extend the driving time just to avoid changing the vehicles);
- time to get from the bus stop to the destination of the trip.

The model of the generalized cost of travel by public transport based on earlier considerations looks as follows (Eqs. 7 and 8):

$$G_{\mathrm{km}} = K_{\mathrm{pr}} + K_{\mathrm{npkm}} \quad (7)$$

where: G_{km}—generalized cost of public transport journey [PLN], K_{pr}—travel costs made by public transport [PLN], K_{npkm}—nonmonetary costs [PLN].

$$K_{\mathrm{npkm}} = \left(t_{\mathrm{do}} + t_{\mathrm{o}} + t_{\mathrm{j}} + t_{\mathrm{prz}} + t_{\mathrm{doc}}\right) \times k \quad (8)$$

where: t_{do}—travel time from starting point to public transport stop [min], t_{o}—waiting time for public transport vehicle [min], t_{j}—journey time (including congestion) [min], t_{prz}—time needed for interchange [min], t_{doc}—travel time to destination point [min], k—unit cost of travel time [zł/h].

Driving time, which is often prolonged due to congestion, can be seen as an advantage, but also as a waste of time. When travelling by public transport, users often use this time to relax, prepare for work, or other activities that they would not be able to do while driving. Therefore, it can be noted that the inevitable stop in the so-called traffic jams can give benefits to public transport passengers. In the case of driving a car (as far as the driver is concerned), this time is used to a lesser extent and the attention to be devoted to travel makes it impossible to perform other activities.

4.1 Generalized Travel Costs

4.1.1 Case Study

Transport users take into account different variants of the service in the selection of means of transport, taking into account elements such as price, quality and travel time. Their influence on decisions can be different. In Szczecin, the features that decide on the choice of public transport are: convenient connections by public transport and a relatively lower cost of travel, while in the case of choosing a passenger car, convenience and too long travel time by means of public transport are indicated.[2] Among the most important features of the functioning of public transport: frequency, punctuality, availability and price were indicated (ordered in relation to the

[2] Data based on authors' research made during the preparation of PhD.

number of indications of a given feature). It should be noted that these features differ in particular cities as well as dynamically change over time and they are conditional and difficult to measure.

While the method of measuring the generalized cost of travel can be universal, the results that can be obtained can be very diverse. It is influenced by, among others: type of travel (distance between the source and the destination), type of car (age, combustion level, brand), type of public transport tickets used (one-time, periodic, the possibility of using statutory discounts). In the case study presented, comparison of generalized travel costs referred primarily to daily travel, carried out regularly (e.g. to work).

Below is an example of calculating the generalized cost of travel in Szczecin, with the following assumptions:

- distance from the source of the trip to its destination:
  - travel starting point: ul. Cynowa 1;
  - travel destination point: plac Armii Krajowej 1;
- car used:
  - Opel Astra K sports tourer (production year—2016);
  - petrol engine: 1.4 turbo (average combustion level: 7 l/100 km);[3]
- public transport ticket used: monthly personal name ticket for all lines;
- the annual subscription for zone A in the paid parking zone used.

These calculations apply only to daily travel from source to destination (work), 5 times a week in 2018[4] (Table 3).

Taking into account the above assumptions, the generalized cost of travel per year for a passenger car is:

$$K_{\text{npos}} = 250 \times \left[(0.08 + 0.66 + 0.17 + 0.17) \times 34.38 \right] = 9282.6\,\text{PLN}$$

$$K_{\text{e}} = 2308.6 + 2500 + 450 + 200 = 5458.6\,\text{PLN}$$

$$G_{\text{so}} = \mathbf{5458.6 + 1500 + 9282.6 = 16,241.2\,PLN}$$

In contrast, the generalized cost of travel per year for public transport is:

$$K_{\text{npkm}} = 250 \times \left[(0.13 + 0.17 + 1 + 0.08 + 0.07) \times 34.38 \right] = 12,462.75\,\text{PLN}$$

$$G_{\text{km}} = \mathbf{(12 \times 162\,PLN) + 12,462.75\,PLN = 14,406.75\,PLN}$$

[3] Average petrol price in 2018: 4.85 PLN.

[4] Working days in 2018—250.

5 Conclusions

The example of Szczecin shows the values of the generalized travel cost in the case of passenger car selection (PLN 16,241.21) and public transport (PLN 14,406.75). The costs of car travel alone are in this particular case higher (PLN 2308.6) than in the case of public transport (PLN 1944). It should be taken into account that the car's operating costs (PLN 5458.6) do not occur in the case of using public transport services. It should also be added that in this case, the cost of credit has not been taken into account, and they may play a big role in the overall cost. The last factor that has an impact on the cost of the generalized are nonmonetary costs expressed in the time necessary for the travel. In the case of a passenger car, it is 65 min a day (which translates into PLN 9282.6 per year), while in public transport it is 87 min a day (PLN 12,462.75). Despite the higher value of nonmonetary costs related to travelling by public transport, the final value of the generalized cost in this case is smaller. It should be noted that the nonmonetary aspect resulting from the travel time is subjective. Everyone has a different approach to it and time associated with travel can also be perceived as an advantage, which is difficult to attribute to cost aspects in this case as well as to be seen as a measurable, financial equivalent loss.

It is worth considering, especially in the face of challenges facing cities in the field of sustainable mobility (which one of the aims is the limitation of passenger cars in daily trips) to create an app or programme that is able to present and help residents calculate the costs of choosing different modes of transport in the city area and benefits (not only financial) that are related to this.

Table 3 Data for the formula for the calculation of the generalized travel cost (average annual values for 2018)

Passenger car		Public transport	
Element of the formula	Value	Element of the formula	Value
K_p[a]	2308.6 PLN/year	K_{pr}	162 PLN/month
K_n	0 PLN	t_{do}[b]	8 min (0.13 h)
K_u	2500 PLN	t_o	10 min (0.17 h)
K_{pt}	450 PLN	t_j	60 min (1 h)
K_k	0 PLN	t_{prz}	5 min (0.08 h)
K_{pz}	200 PLN	t_{doc}	4 min (0.07 h)
K_{park}	1500 PLN	k	34.38 PLN/h
t_{do}	5 min[c] (0.08 h)		
t_j	40 min[d] (0.66 h)		
t_{park}	10 min (0.17 h)		
t_{doc}	10 min (0.17 h)		
K	34.38 PLN/h		

Source: own elaboration
[a]Average two-way distance: 27.2 km
[b]Values based on route calculation programme: jakdojade.pl
[c]Values for time are multiplied by 2 (for return trips)
[d]Including congestion in the morning and afternoon peaks

In the case of measures to increase the use of public transport for everyday travel, it is the responsibility of the city, transport organizers and carriers to improve the offer, or to reduce the generalized costs of users, mainly in the nonfinancial part of costs (saving time, mental effort). Generalized transport costs by collective transport are in the passenger's mind lower, provided that the demand for accessibility and convenience of travel is met (short time to reach the bus stop and waiting for other means of transport, clean and comfortable public transport vehicle, availability of information about the transport service). This cost may be even lower than in the case of using a passenger car, unless the passenger experiences the stress associated with the travel and productively can use the time during the trip (rest, preparation for work, making friends, etc.).

References

1. Załoga, E.: Trendy w transporcie lądowym Unii Europejskiej. Wydawnictwo Naukowe Uniwersytetu Szczecińskiego, Szczecin (2013)
2. Button, K.: Transport Economics. Edward Elgar Publishing Ltd., Great Britain (2010)
3. Button, K.: Transport economics 3rd edition, Edward Elgar, Northampton, MA, Cheltenham UK (2010). In: Załoga, E. (ed.) Trendy w transporcie lądowym Unii Europejskiej, pp. 143–144. Wydawnictwo Naukowe Uniwersytetu Szczecińskiego, Szczecin (2013)
4. Estache, A., de Rus, G.: Privatization and Regulation of Transport Infrastructure. Guidelines for Policymakers and Regulators, p. 163. The World Bank Institute, Washington (2000)
5. Grosso, M.: Variables Influencing Transport Mode Choice: A Generalized Cost Approach, p. 6. Società Italiana di Economia dei Trasporti e della Logistica, Messina. http://www.sietitalia.org/wpsiet/Grosso_Monica_Siet_2011.pdf (2011)
6. Victoria Transport Policy Institute: Transportation Cost and Benefit Analysis II—Travel Time Costs. Victoria Transport Policy Institute, Victoria (2009). http://www.vtpi.org/tca/tca0502.pdf
7. Niebieska Księga: Sektor Transportu Publicznego w miastach, aglomeracjach, regionach. https://www.pois.gov.pl/strony/o-programie/dokumenty/niebieskie-ksiegi-dla-projektow-w-sektorze-transportu-publicznego-infrastruktury-drogowej-oraz-kolejowej/

Challenges Related to the Implementation and Development of Electromobility in Cities

Grzegorz Sierpiński, Ander Pijoan, Katarzyna Turoń, and Marcin Staniek

Abstract Transport is one of the main urban logistics issues. The discussion on the development of transport systems, expressed in consecutive EU White Papers and communications, highlights two main challenges, i.e. the need to change the modal split, including the growing role of public, rail and intermodal transport, as well as changes in energy sources. The response to the second challenge includes, inter alia, electromobility. Technological advancement turned the electric vehicle into a real alternative to the conventional car, especially in urban areas. This enabled to reduce the negative impact of transport on the environment (e.g. reduced emission and energy consumption) which did not require a major increase in the use of public transport.

The chapter describes challenges for the urban logistics as regards the large-scale implementation of electromobility. It distinguishes four main types of activities, such as the development of vehicles, adjustment of infrastructure, improvement of organizational structure and the support of integrated information systems.

Analyses referred to were implemented as a part of the international research project 'Electric travelling—a platform supporting the implementation of electromobility in smart cities based on ICT applications' funded from the National Research and Development Centre under the ERA-NET CoFund Electric Mobility Europe Programme.

Keywords Electromobility · Transport infrastructure · Electric vehicles · ICT

G. Sierpiński (✉) · K. Turoń · M. Staniek
Silesian University of Technology, Gliwice, Poland
e-mail: grzegorz.sierpinski@polsl.pl; katarzyna.turon@polsl.pl; marcin.staniek@polsl.pl

A. Pijoan
DeustoTech - Fundación Deusto Universidad de Deusto, Bilbao, Spain
e-mail: ander.pijoan@deusto.es

M. Suchanek (ed.), *Transport Development Challenges in the 21st Century*, Springer Proceedings in Business and Economics,
https://doi.org/10.1007/978-3-030-50010-8_18

1 Introduction

The transport network is the circulatory system of the city. Movement of people and cargo necessitates relevant organizational setup and also has a negative impact on the city itself. Contemporary challenges forced the improvement of efficiency in companies through the implementation of a new approach to urban logistics. There are several definitions of urban logistics. Most of them make reference to the sustainable urban development (e.g. [1–7]). One of the definitions, which encompasses possibly the largest number of issues, has been presented in [8]. According to the definition, urban logistics is a set of processes, including the management of people, cargo and information flows in the city logistics system, aimed at meeting city development needs and goals. Those needs and goals are met with due respect to the natural environment and the fact that the city is a social organization which overriding goal is to meet users' needs. As regards urban logistics, we may distinguish three types of stakeholders, such as the local council, companies operating in their specific areas of business, and inhabitants and people permanently staying in the city. Negative impact of transport (external cost of transport [9]) can be seen primarily as emission of noxious substances, noise and congestion. From the point of view of the social dimension of a city [10], we should examine choices people make while travelling and their impact on the functioning of a given area [11]. The process can be supported by mobility plans which are introduced to promote specific transportation behaviour [12].

Recently, we have observed a significant development in the area of alternative fuels. It is the consequence of guidelines published, including guidelines for the EU member states [13–15]. According to the guidelines, decarbonization of transport is one of the main global challenges (e.g. [16, 17]). Although electricity has become an alternative to traditional fuel, the development of electromobility necessitates a number of actions and poses a challenge for contemporary cities. Undoubtedly, several social barriers need to be overcome and the same applies to thinking about electric cars. It is, however, a technical issue related to the development of vehicles and transport infrastructure [18, 19], but also information systems:

$$D_{\mathrm{EM}} = f\left(D_{\mathrm{EV}}, D_{\mathrm{TI}}, D_{\mathrm{ICT}} \right) \tag{1}$$

where: D_{EM}—electromobility development, D_{EV}—development of electric vehicles, D_{TI}—development of transport infrastructure, D_{ICT}—development of information systems.

The chapter presents issues related to the implementation of electromobility. Technical changes are combined with information technologies; the latter can substantially shape user behaviour and expedite full implementation of electromobility in urban logistics.

2 Development of Electric Vehicles

Currently, a number of countries have intensified their efforts to develop electromobility. For instance, Poland established the 2016–2025 Electromobility Development Plan [20]. We may expect that in the longer term the technology will prevail in transport, especially that we have already observed an increase in the number of new electric cars registered [21]. According to the Polish national estimates, in 2025, we expect to have one million electric vehicles. According to the EU forecast, we should have eight to nine million electric vehicles by 2020 [13], whereas the global outlook is 20 million [22]. In 2015, the number of electric vehicles (including hybrids) exceeded 1.26 million [22], and in 2016, 2 million [23]. From the point of view of users, major barriers to a wider expansion of those cars are their range and price. According to Table 1, the range of an electric vehicle is insufficient for commuting to work or school. In the case of longer distances, only the most expensive makes can meet expectations. We may compare the positioning of an electric car against other modes. Thus, considering longer distances and the use of public transport (e.g. rail, bus), a larger range does not have to be a rudimentary feature of an electric car.

Unfortunately, the price of an electric car remains the main barrier for users. Figure 1 shows that to buy an electric car one needs to spend PLN100 to PLN160 thousand. Two cheaper makes can meet the range requirement, but due to their limited size they are not able to transport the standard number of passengers.

A number of research projects have been implemented, e.g. Horizon 2020 Programme, as regards the construction of new electric vehicles and improvement in the capacity of their batteries. It should be noted that those projects cover both individual passenger cars and trucks [25]. The latter are an important link in the green supply chain [26] for the first and the last mile [4], especially when a vehicle needs to enter a city centre and due to the emission zoning other types of vehicles are banned [27]. Selected projects of 2015–2018 are listed in Table 2.

The development of electric vehicles does not apply to individual passenger cars only. The recent intensive effort led to the implementation of electric buses in public

Table 1 Electric car ranges. (Source: author's own collaboration based on car manufacturers' websites and [24])

Brand and model	Range[a] [km]	Brand and model	Range[a] [km]
Tesla S	340–540	Hyundai Ioniq Electric	200
Hyundai Kona	390	Renault Kangoo ZE	181
Jaguar I-Pace	354	Ford Focus Electric	185
Nissan Leaf	340	Kia Soul Electric	150
BYD e6	301	e.GO Life	130
Renault Zoe	268	Volkswagen e-up!	107
BMW i3	240	Renault TWIZY	100
Volkswagen e-golf	200	Smart Fortwo Electric Drive	93

[a]for selected makes, longer range corresponds with the most powerful battery

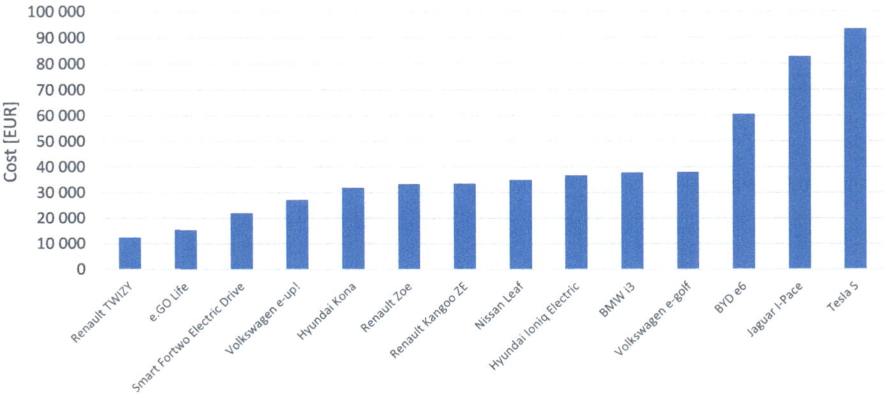

Fig. 1 Electric car prices (Source: author's own collaboration based on car manufacturers' websites and [24])

Table 2 Selected European electromobility research projects (2015–2018). (Source: author's own collaboration based on project's websites and European Commission website [28])

Project acronym	Project full title	Topics	Project's website
EU-LIVE	Efficient Urban LIght VEhicles	Concept and design electric vehicles	eu-live.eu
RESOLVE	Range of Electric SOlutions for L-category VEhicles	Concept and design of electric vehicles	www.resolve-project.eu
ESPRIT	Easily diStributed Personal RapId Transit	Concept and design electric vehicles	www.esprit-transport-system.eu
ECAIMAN	Electrolyte, Cathode and Anode Improvements for Market-near Next-generation Lithium Ion Batteries	Batteries and battery packs	www.ecaiman.eu
SPICY	Silicon and polyanionic chemistries and architectures of Li-ion cell for high energy battery	Batteries and battery packs	www.spicy-project.eu

transport. Apart from public transport advantages, i.e. reduced occupancy of urban space, the development brought a positive influence on the environment by limiting emission [29]. The implementation of electromobility in public transport necessitates a number of organizational changes pertaining to routing and timetable planning. The routing needs to guarantee that a vehicle reaches its charging stations in due time (depot or terminus) (more in: [30–34]).

Apart from public transport and individual cars, urban logistics provides a number of opportunities for organizational and conceptual changes designed to establish and operate electric car and scooter rental [10, 35]. Not only does the urban logistics aim at increasing the number of users of electric vehicles but also integrate various modes of transport. Electric car-sharing can be a good complement to the public transport chain, and incentivise people to use electric vehicles. The idea of combining electric car rental with Park & Ride systems establishes a transport chain, in which people first use their private cars, then public transport, and they cover the last stretch using an environmentally friendly car from a city car rental [36].

3 Changes of Transport Infrastructure

The development of infrastructure involves the building of charging stations. Shortage of such stations is a technical barrier for operating electric vehicles and a mental barrier for users. Locations of such stations should correspond with people's needs. While discussing the issue of electric car charging stations, we need to distinguish three parameters [23]:

- Output power,
- Type of plug,
- Mode of operation defining communication between charger and vehicle.

To promote integration and expedited implementation of electric cars attempts are made to harmonize technical specifications [13]. At the moment, several different types of plugs are used in the world, including type 1, type 2, CHAdeMO, CCS, Commando, 3-pin etc. We can also distinguish slow and fast chargers with rated power below 22 kW and above 22 kW, respectively [23]. Although recently, we witnessed a major increase, differences between specific areas still remain. The diversity in the development of electric car charging stations is presented in Fig. 2. While in certain areas, despite high industrialization (Upper Silesia Conurbation), we can still see a shortage of such transport infrastructure, some cities, e.g. Amsterdam, have become friendly to electric cars [39].

In the nearest future, we can expect a major increase in the number of charging stations. In Poland, until the end of 2020, it is planned to establish at least 1370 charging points [40]. According to the global forecast, until 2025, we expect 4 million slow chargers and 0.1 million fast chargers [23].

4 Support for Integrated Information Systems: Example of Electric Travelling Project

The expected increase in the number of electric vehicles on our roads, the development of infrastructure, and organizational changes has created an information gap regarding travelling options. Therefore, on the one hand, we should look for tools that can facilitate education of users regarding electromobility in an integrated manner, and on the other, such a tool should support local governments in the

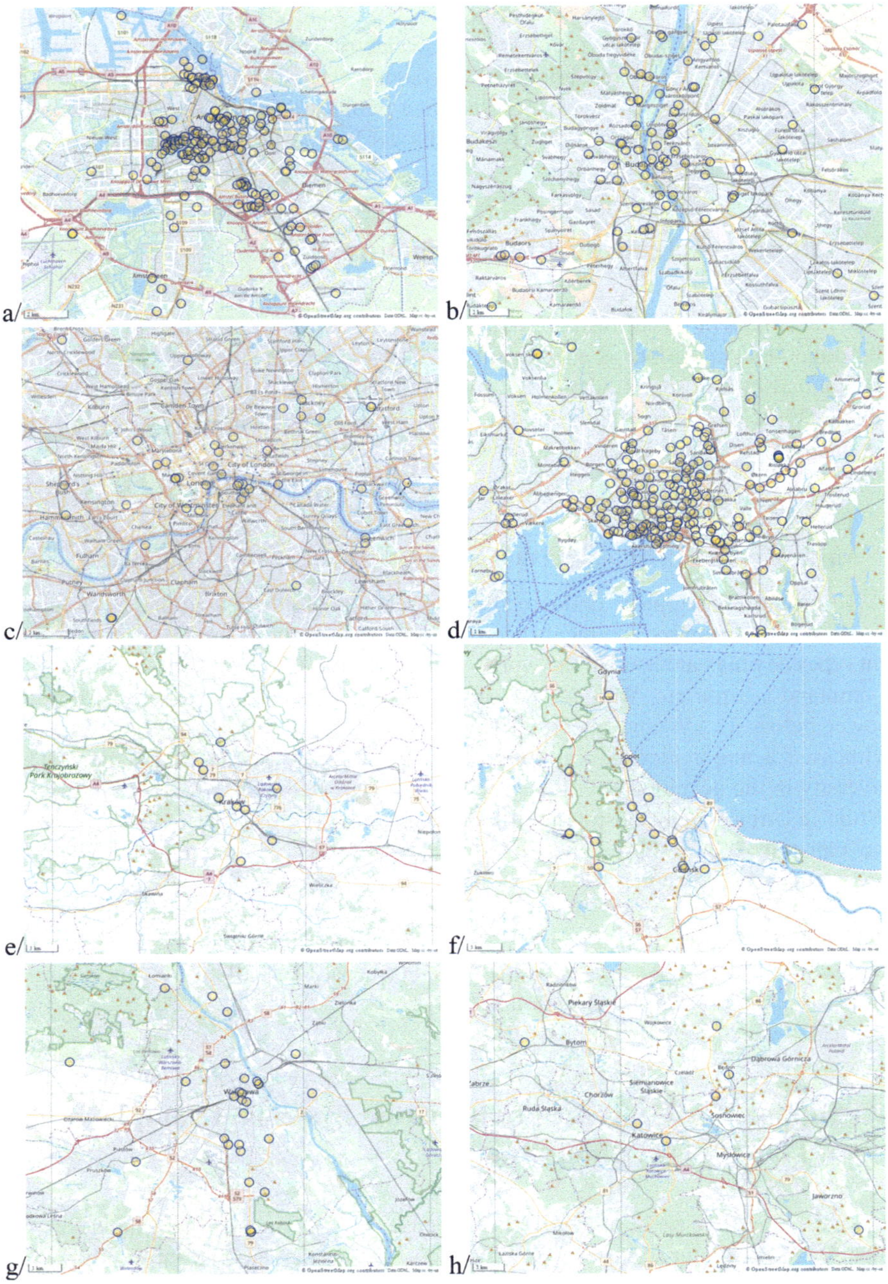

Fig. 2 Location of charging stations in selected cities and areas in Europe: (**a**) Amsterdam, (**b**) Budapest, (**c**) London, (**d**) Oslo, (**e**) Kraków, (**f**) Gdańsk, (**g**) Warszawa, (**h**) Upper Silesia Conurbation (Source: own research based on [37, 38])

implementation of electromobility solutions. The above tool has been developed by the international research project of 'Electric travelling—platform to support the implementation of electromobility in Smart Cities based on ICT applications' (ET Project) [41] under the ERA-NET CoFund Electric Mobility Europe Programme. The analysis of information shortages has helped to identify main needs. The tool developed comprises a trip planner and development scenario simulator.

The trip planner has been designed to provide users with functionalities to choose, visualize and compare possible trips, as well as support the development of the transport network between two (or more) points within a transport chain consisting of one or more means of transport. The majority of trip planners, however, is limited to defining distance and travelling time, and only a few take into account the environment (more in: [42–45]). In the case environmental impact is included. It means user can choose between time, distance but also cost and environmental impact. In all these criteria, optimal solution will be minimum of the value of selected criteria.

Apart from the majority of means of transport in the city, including public transport, individual cars, bikes, and walking trips, the new trip planner (Fig. 3) encompasses transport chains using existing Park & Ride and Bike & Ride systems and car-sharing and bike-sharing systems. Moreover, the planner enables to plan one's trip using an individual electric car, which is particularly important for the promotion of electromobility. The tool is an extension of the route planning system developed by the Green Travelling project [46]. The optimized route is set taking into account charging stations available in the city and the user can receive information about the need to recharge batteries during their trip. This improves the image of the electric car and overcomes the mental barrier that prevents users from buying the electric car due to its limited range. The system, including the planner, can be connected to Smart Transport Systems operating in the city [47–49] and the system can set better routes that take into account congestion and availability of charging stations.

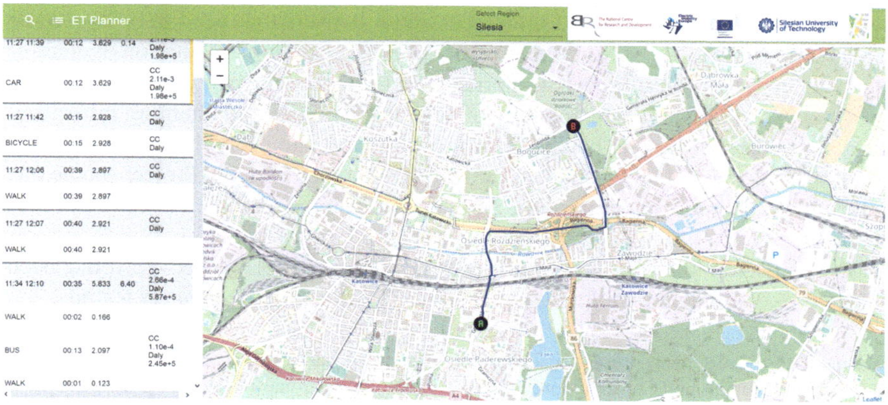

Fig. 3 ETPlanner interface (Source: own research)

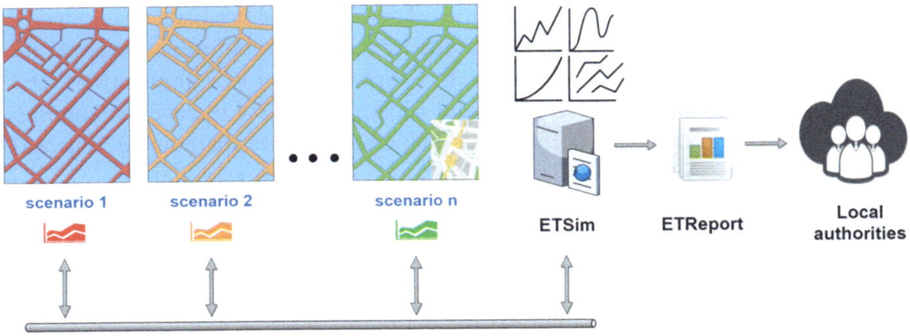

Fig. 4 Multi-agent simulator as support in decision-making process (Source: own research)

The second component of the system is the multi-agent simulator. The tool enables to simulate behaviour of travellers who use various modes of transport [50]. A system-generated report can be used as a support in decision-making [51] and selecting the most appropriate scenario for a specific area (Fig. 4).

Comparisons included in the report collate the current status, locations of new charging stations and selected incentives. The report provides suggestions for the local government to facilitate planning of electromobility and expedite the introduction of electric vehicles in the city. Since the planner is based on a heuristic approach, the local government can introduce limitations to the traffic for specific groups of users directly through a routing algorithm [52]. Real needs of travelling persons can be collected based on user queries recorded in Big Data server.

5 Summary

Environmental benefits and incentives for users make electric cars more popular in the world. However, we should remember about barriers that need to be overcome on the way to make the solution really effective. The full implementation of electromobility requires a multifaceted approach and integration with existing transport systems. Apart from the development of vehicles, batteries and charging infrastructure, users' behaviour can be changed by well-tuned and properly delivered information. As discussed in the chapter, the tool developed by the ET Project can be one of elements supporting promotion and faster implementation of electromobility in cities.

It is expected that further research will develop the tool. Moreover, it has been planned to combine the tool with the existing ITS in a specific area for testing.

Acknowledgements The present research has been financed from the means of the National Centre for Research and Development as a part of the international project within the scope of ERA-NET CoFund Electric Mobility Europe Programme 'Electric travelling—platform to support the implementation of electromobility in Smart Cities based on ICT applications'.

The National Centre for Research and Development · Electric Mobility Europe · European Commission · Silesian University of Technology

References

1. Banister, D.: The sustainable mobility paradigm. Transp. Policy. **15**, 73–80 (2008)
2. Jacyna, M., Żak, J., Jacyna-Gołda, I., Merkisz, J., Merkisz-Guranowska, A., Pielucha, J.: Selected aspects of the model of proecological transport system. J. Kones. **20**, 193–202 (2013)
3. Kauf, S.: City logistics—a strategic element of sustainable urban development. Transp. Res. Procedia. **16**, 158–164 (2016)
4. Mindur, L. (ed.): Technologie transportowe. Wydawnictwo Naukowe Instytutu Technologii Eksploatacji – PIB, Radom (2014)
5. Mindur, M.: Logistyka. Nauka – Badania – Rozwój. Wydawnictwo Naukowe Instytutu Technologii Eksploatacji – PIB, Radom (2017)
6. Stanley, J.: Land use/transport integration: starting at the right place. Res. Transp. Econ. **48**, 381–388 (2014)
7. Taniguchi, E.: Concepts of city logistics for sustainable and liveable cities. Procedia Soc. Behav. Sci. **151**, 310–317 (2014)
8. Szołtysek, J.: Podstawy logistyki miejskiej. Wydawnictwo Akademii Ekonomicznej, Katowice (2007)
9. Pawłowska, B.: Zrównoważony rozwój transportu na tle współczesnych procesów społeczno-gospodarczych. Wydawnictwo Uniwersytetu Gdańskiego, Gdańsk (2013)
10. Szołtysek, J.: Ekonomia współdzielenia a logistyka miasta—rozważania o związkach. Gospodarka Materiałowa i Logistyka. **11**, 2–9 (2016)
11. Hebel, K.: Zachowania transportowe mieszkańców w kształtowaniu transportu miejskiego. Wydawnictwo Uniwersytetu Gdańskiego, Gdańsk (2013)
12. Starowicz, W.: Zarządzanie mobilnością wyzwaniem polskich miast. Transp. Miejski Region. **1**, 42–47 (2011)
13. Communication from the Commission to the European Parliament, The Council, The European Economic and Social Committee and the Committee of the Regions: Clean Power for Transport: A European Alternative Fuels Strategy, p. 17. COM, Brussels (2013)
14. White paper on the future of Europe, reflections and scenarios for the EU27 by 2025, COM 2025. (2017)
15. White Paper: roadmap to a single European transport area—towards a competitive and resource efficient transport system. COM 144 (2011)
16. Santos, G.: Road transport and CO2 emissions: what are the challenges? Transp. Policy. **59**, 71–74 (2017)
17. van der Zwaan, B., Keppo, I., Johnsson, F.: How to decarbonizes the transport sector? Energ Policy. **61**, 562–573 (2013)
18. Gajewski, J., Paprocki, W., Pieriegud, J. (red.): E-mobilność: wizje i scenariusze rozwoju. Centrum Myśli Strategicznych, Sopot (2017)
19. Grzelakowski, A.S., Matczak, M. (eds.): Infrastruktura transportu. Współczesne wyzwania rozwojowe. Wydawnictwo IMP PAN, Gdańsk (2015)
20. Energia do przyszłości: Plan rozwoju elektromobilności w Polsce. Ministerstwo Energii, Warszawa (2016)
21. European Vehicle Market Statistics. Pocketbook 2017/18. http://eupocketbook.theicct.org. Accessed 20 July 2018
22. Global EV Outlook. Beyond one million electric cars. Int. Energ. Agency. **2016** (2016)
23. Global EV Outlook.Two millions and counting. Int. Energ. Agency. **2017** (2017)

24. Elektrowoz service. www.elektrowoz.pl (30.01.2019)
25. Wątróbski, J., Małecki, K., Kijewska, K., Iwan, S., Karczmarczyk, A., Thompson, R.G.: Multi-criteria analysis of electric vans for city logistics. Sustainability. **9**(8), 1453 (2017)
26. Tundys, B.: The impact and role of transportation on the construction and operations of the green supply chain. In: Suchanek, M. (ed.) Sustainable Transport Development, Innovation and Technology, Springer Proceedings in Business and Economics, 15-26. Springer, Cham (2017)
27. Dembińska, I., Jedliński, M., Marzantowicz, Ł.: Logistic support for a rescue operation in the aspect of minimizing the ecological footprint as an environmental requirement within sustainable development on the example of a natural disaster. LogForum. **14**(3), 355–370 (2018)
28. European Commission website. https://ec.europa.eu/commission/index_en
29. Krawiec, S., Łazarz, B., Markusik, S., Karoń, G., Sierpiński, G., Krawiec, K., Janecki, R.: Urban public transport with the use of electric buses—development tendencies. Transp. Probl. **11**(4), 127–137 (2016)
30. Celiński, I., Sierpiński, G.: Real time model for public transportation management. LogForum. **10**(1), 31–41 (2014)
31. Grzelec, K., Birr, K.: Development of trolleybus public transport in Gdynia as part of a sustainable mobility strategy. Sci. J. Sil. Univ. Technol. Series Transp. **92**, 53–63 (2016)
32. Lejda, K., Mądziel, M., Siedlecka, S., Zielińska, E.: The future of public transport in light of solutions for sustainable transport development. Sci. J. Sil. Univ. Technol. Series Transp. **95**, 97–108 (2017)
33. Markusik, S., Krawiec, S., Łazarz, B., Karoń, G., Janecki, R., Sierpiński, G., Krawiec, K.: The technical and operational aspects of the introduction of electric—powered buses to the public transportation system. Logist. Transp. **27**(3), 41–50 (2015)
34. Pawlak, Z., Stajniak, M.: Optimisation of transport processes in city logistics. LogForum. **7**(1), 15–22 (2011)
35. Ocicka, B., Wieteska, G.: Sharing economy in logistics and supply chain management. LogForum. **13**(2), 183–193 (2017)
36. Sierpiński, G., Turoń, K., Pypno, C.: Urban transport integration using automated garages in Park&Ride and car-sharing systems—preliminary study for the upper Silesian conurbation. In: Sierpiński, G. (ed.) Integration as Solution for Advanced Smart Urban Transport Systems, Advances in Intelligent Systems and Computing, vol. 844, pp. 218–228. Springer, Cham (2019)
37. OpenStreetMap service. https://www.openstreetmap.org/
38. Overpass service. http://overpass-turbo.eu/
39. Roman, D., Czech, P., Sierpiński, G., Turoń, K., Urbańczyk, R.: Amsterdam jako miasto przyjazne samochodom osobowym z napędem hybrydowym i elektrycznym. Autobusy. Techn. Eksploat. Syst. Transp. **6**, 396–404 (2017)
40. Ustawa z dnia 11 stycznia 2018 r. o elektromobilności i paliwach alternatywnych. Dz. U. z 2018 r. poz. 317, 1356 (2018)
41. Electric travelling—platform to support the implementation of electromobility in Smart Cities based on ICT applications—Project proposal under EMEurope programme (2016)
42. Borkowski, P.: Towards an optimal multimodal travel planner—lessons from the European experience. In: Sierpiński, G. (ed.) Intelligent Transport Systems and Travel Behavior, Advances in Intelligent Systems and Computing, vol. 505, pp. 163–174. Springer, Cham (2017)
43. Esztergár-Kiss, D., Csiszár, C.: Evaluation of multimodal journey planners and definition of service levels. Int. J. Intell. Transp. Syst. Res. **13**, 154–165 (2015)
44. Lewczuk, K., Żak, J., Pyza, D., Jacyna-Gołda, I.: Vehicle routing in urban area—environmental and technological determinants. Urban Transp. XIX. **130**, 373–384 (2013)
45. Sierpiński, G.: Technologically advanced and responsible travel planning assisted by GT planner. In: Macioszek, E., Sierpiński, G. (eds.) Contemporary Challenges of Transport Systems

and Traffic Engineering, Lecture Notes in Network and Systems, vol. 2, pp. 65–77. Springer, Cham (2017)
46. Sierpiński, G.: Using ICT applications to support sustainable development and tackle the barriers related to extensive introduction of electric cars into transport systems—a case study based on the green travelling planner tool. In: Suchanek, M. (ed.) New Research Trends in Transport Sustainability and Innovation: TranSopot 2017 Conference. Springer Proceedings in Business and Economics, pp. 62–72. Springer, Cham (2018)
47. Cárdenas, O., Valencia, A., Montt, C.: Congestion minimization through sustainable traffic management. A micro-simulation approach. LogForum. **14**(1), 21–31 (2018)
48. Iwan, S., Małecki, K.: Data flows in an integrated urban freight transport telematic system. In: Mikulski, J. (ed.) Telematics in the Transport Environment. TST, Communications in Computer and Information Science, vol. 329, pp. 79–86. Springer, Berlin (2012)
49. Szymczak, M., Sienkiewicz-Małyjurek, K.: Information in the city traffic management system. The analysis of the use of information sources and the assessment in terms of their usefulness for city routes users. LogForum. **7**(2), 37–50 (2011)
50. Pijoan, A., Kamara-Esteban, O., Alonso-Vicario, A., Borges, C.E.: Transport choice modeling for the evaluation of new transport policies. Sustainability. **10**, 1230 (2018)
51. Galińska, B.: Intelligent decision making in transport. Evaluation of transportation modes (types of vehicles) based on multiple criteria methodology. In: Sierpiński, G. (ed.) Integration as Solution for Advanced Smart Urban Transport Systems, Advances in Intelligent Systems and Computing, vol. 844, pp. 161–172. Springer, Cham (2019)
52. Sierpiński, G., Staniek, M.: Heuristic approach in a multimodal travel planner to support local authorities in urban traffic management. Transp. Res. Procedia. **27**, 640–647 (2017)

Fare-Free Public Transport vs. Private Cars: Zero Fares as an Instrument of Impact on Public Transport Mode Share?

Anna Urbanek

Abstract For several years in Europe, and particularly over the last few years in Poland, there has been a lively discussion about fare-free public transport (FFPT). The discussions focus mostly on the effectiveness of urban public transport fares in increasing the attractiveness and competitiveness of public transport in cities. This is particularly important in the era of increasing congestion and deteriorating quality of life in cities. The price, after all, directly affects the market demand; it is also the most powerful instrument of marketing mix influencing the customer behaviour. Moreover, public transport fares are an essential tool of transport policy. The aim of the chapter is to investigate the effectiveness of FFPT in increasing its competitiveness, especially in relation to private cars. The chapter presents the results of a survey carried out among individuals commuting by private car in the central part of Silesia Province in Poland.

Keywords Public transport · Fare-free · Fareless · Modal shift · Travel behaviour · Private cars · Travel attitudes · Sustainability · Zero fares

1 Introduction

The dynamic development of road transport, including especially private car traffic that we have been experiencing globally for over 40 years, is one of the major sources of natural environment pollution, noise, as well as increasing problems related to congestion and deteriorating quality of life for inhabitants of cities. A vast majority of journeys begin and end in cities, and it is precisely the cities that are most affected by the negative effects of modal imbalance in passenger transport. In

A. Urbanek (✉)
Department of Transport, University of Economics in Katowice, Katowice, Poland
e-mail: anna.urbanek@ue.katowice.pl

M. Suchanek (ed.), *Transport Development Challenges in the 21st Century*, Springer Proceedings in Business and Economics,
https://doi.org/10.1007/978-3-030-50010-8_19

cities around the world, it is necessary to undertake actions that will result in reducing private car traffic in favour of alternative, more resource-saving and environmentally friendly transport modes, especially actions towards increasing public transport share in fulfilling the mobility needs of inhabitants of cities.

Over the last 10–15 years, the world literature on the subject has been increasingly focused on the issues related to the efficiency of the impact of fare-free public transport (FFPT). In recent years, this issue has also been increasingly undertaken in the literature on transport economics worldwide, as well as in Poland [1–4]. Lively discussions are held as part of the public and scientific agenda regarding the efficiency of FFPT as an instrument for increasing the attractiveness and competitiveness of public transport in comparison with individual car transport.

In this context, what appears to be particularly significant is the knowledge about the attitudes of private car users towards public transport, as well as the knowledge about the extent to which travel cost is the decisive factor for selection of a specific means of transport. To obtain this knowledge is the main purpose of the conducted survey, the selected results of which were presented in this article.

The main aim of the article is to present the results of the survey conducted among individuals commuting by private car and to discuss the previous research results in respect of the efficiency of impact of fare-free public transport on the increase of public transport mode share.

2 Fare-Free Public Transport and Modal Shift: A Literature Review

2.1 The Effect of Fares on Public Transport Demand

The sensitivity of passengers to changes in public transport services is a relatively well-researched issue in the world literature on the subject. There are available research results from different countries, conducted for various market conditions, e.g. Webster and Bly [5]; Oum et al. [6]; Goodwin [7]; Paulley et al. [8]; de Rus [9]; Dargay and Hanly [10]; Lago et al. [11], Hensher [12], Cervero [13].

The results of previously conducted research indicate that passengers are more sensitive to changes of transport offer and quality of services than to changes of prices. Changes in transport offer, i.e. reduction of travel time and increase of frequency have a better chance of positive impact on increasing the number of passengers in public transport than reduction of fares [13–15]. This may be confirmed by the results of meta-analysis of public transport demand conducted by Holmgren [16], who performed an analysis and summarised the previously conducted research on bus demand elasticities (Table 1).

Elasticity of demand can be defined as the ratio of relative (percentage) change of demand volume and relative (percentage) change of factor affecting the demand. Negative value means changes in the opposite direction, i.e. for example an increase of a specific factor implies a decrease of demand. Positive value means changes in this direction [16, 17].

Table 1 Elasticity values from the bus demand studies included in the meta-analysis performed by Holmgren (Source: [16])

Variable	Number of studies	Mean	Std. error	Min	Max
E_p	81	−0.38	0.23	−1.32	−0.009
E_{vkm}	58	0.72	0.37	0.075	1.88

E_p fare elasticity of demand, E_{vkm} elasticity with respect to vehicle-kilometres supplied

As follows from the data gathered in Table 1, the average fare elasticity of bus demand arising from the research is lower (absolute value) than the average value for elasticity with respect to the vehicle-kilometres supplied. This means that an increase of operational work has a stronger impact on the increase of the number of passengers than reduction of prices.

Price elasticity can also be calculated in case of abolishing fares for public transport services and may be used for forecasting change in demand following the introduction of FFPT. It should be calculated on the basis of mid-point arc elasticity formulation [13, 17]. However, some researchers have drawn attention to the fact that the suitability of traditional fare elasticity is limited in case of introducing FFPT [18].

There are many factors that determine the sensitivity of passengers to changes of public transport prices. The level of this sensitivity depends e.g. on age, income, professional activity, availability and prices of alternative travel modes, as well as on the geographic features of the examined area [17, 19].

The previously conducted research has also demonstrated that the sensitivity of passengers to PT fares depends on the quality of transport offer. The research by Lago et al. [20] conducted in big cities demonstrates that the passengers of rapid rail service were significantly less sensitive to prices than bus passengers, which probably resulted from better quality of services, especially in respect of time travel and reliability.

According to Hensher et al. prices are the most important factor that is ultimately decisive for the perception of public transport, whereas the quality of services is only taken into consideration by the users as a second choice [21]. However, most importantly, Hensher et al. pays attention to the fact that the price is always referred to the quality received by the user and assessed only in the context of what the passenger buys for a specific price [21]. Therefore, the solution is not to lower prices, but to improve quality, so that it corresponds to the service price.

2.2 *Effectiveness of Fare Abolition on Public Transport Mode Share*

Complete abolition of public transport fares, i.e. implementation of the so-called 'zero fare' rarely concerns the entire public transport system in a specific area, e.g. in a city. This solution is also relatively rarely implemented over a long period of time, e.g. above 1 year. It much more frequently concerns a selected part of a city, selected routes, specific social or age groups, distinguished times of the day or selected days in a year. For example, in Europe, free public transport is relatively commonly introduced on the occasion of Car-Free Day (celebrated on 22nd September) for all passengers, or only for the owners of private cars. In some cities, fare abolition is extended for the entire European Mobility Week. Abolition of fares for public transport services is sometimes the effect of dramatic events, such as e.g. terrorist attacks, natural disasters, or the days when air pollution limits are significantly exceeded, which has become quite a common practice in Poland in recent years [22].

In cities around the world, zero fare is most frequently applied only to a limited extent, e.g. for selected passenger groups. Children until certain age, people with various degrees of disability, elderly people after certain age and low-income groups are very often entitled to free journeys.

FFPT is a tool whose use is intended to contribute to achieving specific goals. Based on the cities which abolished public transport fares in whole or in part, the following three main groups of goals of introducing FFPT can be distinguished [15, 23, 24]:

- modal shift from car to public transport: reducing car dependency and car externalities (especially air pollution and noise),
- increased availability of public transport for elderly people, students, low-income groups or the unemployed,
- efficiency goals: reduction of congestion costs in cities and reduction of expenditure on road transport infrastructure.

As noticed by Hebel et al., regardless of the presented objectives, the introduction of FFPT always has a certain political dimension, which may be evidenced by intensification of activities and discussions in this respect during electoral campaigns [3].

Over the last 10–15 years, the world literature on the subject has been increasingly focused on the issues related to the efficiency of the impact of FFPT on the increase of public transport mode share. However, the research on this issue is not a novelty in transport economics. We may refer to the research conducted in 1960s and 1970s [25]. It is worth mentioning very interesting results of the survey conducted in 1962–1963 in Chicago by Moses and Williamson [25]. Back then, only 13% of car users declared that they would switch to public transport, if it was free. The interesting thing is that in order to convince 30% of respondents to switch to public transport, each of them would have to be paid extra 15 cents for each journey, whereas 47% of respondents would require as much as 30 cents to be paid for each journey [25]. Therefore, the prices of public transport services did not constitute a strong factor motivating to switch from private car.

The analysis of the survey carried out on the individuals commuting to Brussels, conducted in 2005 by De Witte et al., leads to similar conclusions [26]. The survey was carried out on 1276 respondents (536 private car users and 740 railway passengers). According to the conducted research, only 10% of respondents commuting by private car declared the willingness to switch and use free public transport;

42% are the respondents who were unable to declare such willingness unambiguously, whereas as much as 48% of respondents would definitely not change the transport mode. The following obstacles that discourage the respondents from using public transport in daily commuting were most frequently mentioned: lack of suitable connections (61%) and insufficient travel speed (46%). If the main obstacles were removed, as much as 25% of private car users declared the willingness to use public transport instead of private cars. 22% of private car users answered negatively [26].

In the source literature, there are also available research results concerning the analysis of actual behaviour changes after introducing fare-free public transport. Fare-free public transport was introduced in 1997 in Hasselt (a town in Belgium with approximately 72,000 inhabitants). The number of passengers in public transport increased four times (from 360,000 to almost 1.5 million) already during the first year of the operation of fare-free public transport [27]. In the subsequent years, the demand for urban transport was constantly increasing. In 1997–2012, the number of bus passengers increased tenfold [28]. The increasing number of passengers and growing expectations regarding the transport offer caused the necessity of more than tenfold increase of the expenditure on public transport, which consequently forced the authorities of Hasselt in 2013 to withdraw from the zero fare and introduce transport fares. The entitlement to fare-free public transport was maintained only for the group of young people aged up to 19 [27]. However, the research on the so-called additional passengers in Hasselt demonstrated that only 16% of them were individuals who would use a private car if public transport was not free, whereas the others were the substitution effect of pedestrian movements and bike journeys (20% of additional passengers), as well as additional mobility—as many as 60% of additional passengers declared that if public transport was free, they would not need to travel [28].

As indicated by Boussauw and Vanoutrive, a similar phenomenon was noted in a group of students in Leuven, who were exempted from public transport fares between 2001 and 2013. It turned out that a vast majority of new public transport passengers were students who had previously travelled to university by bike [29].

The research conducted so far demonstrates that the assumed goal of shifting a part of private car movement to public transport was also not achieved in Tallinn, where FFPT was introduced in 2013. To date, Tallinn is the biggest city in the world that introduced fare-free public transport for all its inhabitants (approximately 420,000 of inhabitants). Before the introduction of zero fares, Tallinn was among the cities with relatively sustainable distribution of transport tasks and relatively low level of public transport prices [30], because the public transport share in fulfilling transport needs in the city was relatively high—it amounted to approximately 40%. The research conducted by Cats et al. demonstrates that the introduction of zero fare in a period of less than 1 year contributed to the increase of public transport share in distribution of transport tasks in the city by 8 percentage points (from 53% to 63%), decrease of car journeys amounts to 3 percentage points (from 31% to 28%), whereas pedestrian movements decreased from 12% to 7%, and the share of bike journeys (1%) remained unchanged [30].

The results of research conducted by Thogersen [31], Thogersen and Moller [32], DeWitte et al. [33] indicate that the increased number of public transport journeys can only be observed during the initial period of fare abolition; this phenomenon is rather short-lived and very hard to maintain.

Therefore, the previous research results show that the introduction of fare-free public transport did not lead to a significant and durable modal shift from private cars to public transport and, consequently, to reduction of passenger traffic and congestion in cities, but it caused to a greater extent a substitution of movements unfavourable from the perspective of sustainable development [24].

3 Results of Empirical Studies in Upper Silesia Region in Poland

3.1 Methods

The survey was conducted among the individuals commuting by private car in the central part of Silesia Province in Poland, in the municipalities that jointly organise public transport (within the framework of Municipal Transport Union of the Upper Silesian Industrial District—KZK GOP). It is a highly urbanised area, inhabited by approximately two million people, where the increase of private car share with simultaneous decrease of public transport share in model split has been observed for many years.

The research was conducted in the period from 15 November to 20 December 2017 using two methods: CAPI (Computer Assisted Personal Interview) and CAWI (Computer Assisted Web Interviewing). The study comprised approximately 900 respondents; however, only 343 completed surveys were received. Ultimately, due to formal reasons, 317 surveys were subject to analysis.

The employees of the companies drawn to take part in the research who use their own private car in daily commuting took part in the survey. The companies chosen for the research were drawn using the stratified sampling method. Ten cities were drawn in the first place, out of 29 municipalities organised in KZK GOP (Municipal Transport Union of the Upper Silesian Industrial District). The second stage consisted in randomly choosing one company in every city for the survey from the database of medium-sized and large companies (overall 10 companies). The sample involved both public and private companies, including five large companies and five medium-sized companies. A similar number of survey questionnaires were completed in each company (30–33 correctly completed questionnaires in each company).

The majority of the respondents were men (54.3%). The most numerous group were people aged 35–44 (34.7%). People aged 45–54 were 22.1% of respondents, whereas people aged 25–34 were 19.5% of respondents. The others included elderly people, younger people or individuals who did not declared their age. The most

Table 2 Characteristics of survey respondents with respect to declared net income per family member and car ownership. Source: Own study

Declared net income per family member (PLN)	**Up to 1000**	**Up to 2000**	**Up to 3000**	**Above 3000**	**No response**
Responses [%]	11.7	40.1	17.9	7.6	22.7
Number of cars available in the household (average 1.51)	**1**	**2**	**3**	**4**	**5**
Responses [%]	58.3	34.4	5.4	1.3	0.6

numerously represented were people with secondary education (58.1%) or higher education (30.9%). One private car was available in a majority of households that the respondents were members of (Table 2).

40.1% of respondents who answered the question concerning the income declared the net income per family member in the amount of up to PLN 2000. It is worth adding that the average monthly disposable income per one person in Silesia Province in 2017 amounted to PLN 1645.75.

3.2 Results

The dominating reasons for selection of private car for daily commuting to work included shorter travel time (25.8% of responses) and travel comfort (16.7%). Other causes, such as e.g. necessity to transport family members, e.g. children, lack of necessity to make transfers or lack of necessity to wait for public transport, were indicated with similar frequency at the level of 4.9–8.8%. Only 5.3% of respondents indicated the lack of alternative connection by public transport as the reason for selection of private car.

The respondents were also asked to answer the question regarding the standard average travel time by private car (understood as the door-to-door time). This was followed by the question in which the respondents were requested to state travel time in case of using public transport, also understood as the entire travel time, i.e. door-to-door. Both responses were compared at the level of every respondent, and then aggregated (Table 3).

The interesting fact is that for 4.7% of respondents commuting by public transport would take exactly the same time as commuting by private car, or even 10 min shorter (1.3%). For 22.3% of respondents this time would be only 15 min longer, whereas for further 27.4% the travel time would be extended by between 20 and maximum 30 min.

The respondents were also asked about the average monthly expenses on fuel for their private car that they use for commuting to work. A vast majority of the respondents spend PLN 200–500 (45.5% of answers) or PLN 500–1000 (30.9%) on fuel. The respondents were also asked to indicate the level of average monthly expenses on fuel that would force them to switch from private car use in daily commuting. Again, both responses were compared for each respondent and aggregated (Table 4).

Table 3 Declared commuting time by public transport compared to commuting time by private car ($n = 317$). Source: Own study

Answer	Responses [%]
10 min. shorter	1.3
0 min. (the same time)	4.7
5 min. longer	3.2
10 min. longer	8.5
15 min. longer	10.7
20–30 min. longer	27.5
30–40 min. longer	7.9
40–60 min. longer	10.7
1 h longer or more	12.6
More than 1.5 h longer	2.5
I don't know	5.0
There is no such connection	5.4

Table 4 Declared willingness to pay more for fuel in a month on the average ($n = 317$). Source: Own study

Difference between the current level of expenditure on fuel and the level declared as maximum [PLN] (I am willing to pay more by….)	Responses [%]
0 (I am not able to spend more)	2.2
300 more	0.9
500 more	26.8
800 more	4.7
1000 more	33.8
1300 more	0.3
1500 more	11.7
2000 more	1.9
2500 more	0.9
3300 more	0.3
The amount of expenditure on fuel is not important for me	9.2
I don't know or lack of response	7.3

The respondents were characterised by relatively small sensitivity to increase of expenditure on fuel. Over 33.8% of respondents are willing to pay for fuel as much as PLN 1000 more compared to the current level of expenses. 11.7% of respondents declare that they are willing to pay even up to PLN 1500 more. Moreover, 37.9% of respondents admitted that they were willing to spend on fuel even up to maximum PLN 1500 per month, which means that more than every third respondent is willing to spend on fuel one third of the average monthly gross remuneration in Silesia Province, which in 2017 amounted to PLN 4481.57.

Interestingly, only less than a half of the respondents (48.6%) would use public transport if they were forced to it e.g. due to high prices of fuel. As many as 37.8% of respondents declared that they would still commute by private car, but they would decide to share journeys with other individuals. 10.1% of respondents would decide to commute to work by bike, whereas 3.5% would choose walking.

Among the most important factors that constitute obstacles in public transport use in daily commuting for the respondents include those connected with the transport offer: lack of proper connections, too long travel time and insufficient frequency. These results are convergent with the research results from the Brussels area obtained by De Witte et al. [26]. However, the factors most frequently mentioned as such that would encourage the respondents to use public transport to the largest extent include those directly related to the difficulties with car use, i.e. increase of fuel prices, as well as increase or introduction of parking charges in the workplace rather than those related to public transport offer and quality.

Therefore, the obtained results show that the respondents are willing to bear very high expenses related to the use of the means of transport that fulfil their needs in daily commuting to work best. The obtained results show how respondents actually evaluate shorter travel time, comfort and quality features that travelling in a private car offers them. However, it is worth noting that the car is something more than just a transport mode. The researchers dealing with this subject draw attention to the fact that psychological factors are very important in case of private cars. Car is presently a tool for expressing personality and a symbol of social status [34]. Moreover, as indicated in research conducted both in 1990s and nowadays, travelling by car has been constantly identified with the sense of freedom and independence [35–37].

4 Discussion and Conclusions

The issues related to modal shift from private cars to public transport are a constantly current and important research problem of large significance not only for the development of the theory of transport economics, but also for the business practice.

The previous research results show that price tools in the form of ticket prices or fare-free public transport have limited significance and low efficiency in stimulating switch from private car. The examples of previous implementation of FFPT do not provide unambiguous evidence for the efficiency of impact on significant and durable modal shift from private cars to PT.

The results of the presented surveys show that the sensitivity to commuting costs is low. The respondents are willing to bear very high expenses related to use of own private cars for daily commuting to work in return for shorter travel time and quality of travel thanks to having their own car. Moreover, more than a half of the examined users of private cars would use a different option than public transport if they were forced to switch from private car use.

The results of conducted research show that the increase of public transport share in serving the needs related to daily commuting to workplace and school requires a complex approach. The impact of public transport fares is not sufficient for achieving goals related to modal shift from private cars to PT. First of all, the aim should be to improve the public transport offer through the development of proper network and frequency of connections. However, due to the large significance of psychological factors, at the same time it is necessary to undertake actions towards increasing the inconvenience of car use in cities, e.g. through proper cost management and availability of parking spaces.

References

1. Bąkowski, H.: Economic attitude to the issue of zero tariff in the urban transport. Public Transp. **1**, 25–29 (2018). http://komunikacja.kzkgop.com.pl/public_media/fb/files/KP1_ANG_mail_OK.pdf
2. Tomanek, R.: Free-fare public transport in the concept of sustainable urban mobility. Transp. Probl. **12**, 95–105 (2017)
3. Hebel, K., Wołek, M., Jagiełło, A.: Trends in free fare transport on the urban transport service market in Poland. In: Sierpiński, G. (ed.) Integration as Solution For Advanced Smart Urban Transport Systems. TSTP 2018, Advances in Intelligent Systems and Computing, vol. 844, pp. 208–2017. Springer, Cham (2019)
4. Grzelec, K.: Bezpłatna komunikacja miejska - cele oraz uwarunkowania jej wprowadzenia i funkcjonowania. Transp. Miejski Regional. **4**, 4–11 (2013)
5. Webster, F.V., Bly, P.H. (eds.): The Demand for Public Transport. Report of an International Collaborative Study. Transport and Road Research Laboratory, Crowthorne, Berkshire (1980)
6. Oum, T.H., Waters II, W.G., Yong, J.-S.: Concepts of price elasticities of transport demand and recent empirical estimates. An interpretative survey. J. Transp. Econom. Policy. **26**, 139–154 (1992)
7. Goodwin, P.B.: A review of new demand elasticities with special reference to short and long run effects of price changes. J. Transp. Econom. Policy. **26**, 155–170 (1992)
8. Paulley, N., Balcombe, R., Mackett, R., Titheridge, H., Preston, J., Wardman, M., Shires, J., White, P.: The demand for public transport: The effects of fares, quality of service, income and car ownership. Transp. Policy. **13**, 295–306 (2006)
9. De Rus, G.: Public transport demand elasticities in Spain. J. Transp. Econom. Policy. **24**(2), 189–201 (1990)
10. Dargay, J.M., Hanly, M.: The demand for local bus services in England. J. Transp. Econom. Policy. **36**(1), 73–91 (2002)
11. Lago, A.M., Mayworm, P., McEnroe, J.M.: Transit service elasticities: evidence from demonstrations and demand models. J. Transp. Econom. Policy. **15**(2), 99–119 (1981)
12. Hensher, D.A.: Establishing a fare elasticity regime for urban passenger transport. J. Transp. Econom. Policy. **32**(2), 221–246 (1998)
13. Cervero, R.: Transit pricing research: a review and synthesis. Transportation. **17**(2), 117–139 (1990)
14. Hess, D.B., Yoh, A., Iseki, H., Taylor, B.: Increasing transit ridership: a survey of successful transit systems in the 1990. J. Public Transp. **5**(3), 33–66 (2002)
15. Hess, D.B.: Decrypting fare-free public transport in Tallinn, Estonia. Case Stud. Transp. Policy. **5**, 690–698 (2017)
16. Holmgren, J.: Meta-analysis of public transport demand. Transp. Res. A Policy Pract. **41**(10), 1021–1035 (2017)
17. Litman, T.: Understanding Transport Demands and Elasticities. How prices and other factors affect travel behavior. Victoria Transport Policy Institute, Victoria (2017). http://www.vtpi.org/elasticities.pdf
18. Chen, C., Varley, D., Chen, J.: What affects transit ridership? A dynamic analysis involving multiples factors, lags and asymmetric behaviour. Urban Stud. **48**(9), 1893–1908 (2011)

19. De Grange, L., González, F., Muñoz, J.C., Troncoso, R.: Aggregate estimation of the price elasticity of demand for public transport in integrated fare systems: the case of Transantiago. Transp. Policy. **29**, 178–185 (2013)
20. Lago, A.M., Mayworm, P., McEnroe, J.M.: Further evidence on aggregate and disaggregate transit fare elasticities. Transp. Res. Record. **799**, 42–47 (1981)
21. Hensher, D.A., Stopher, P., Bullock, P.: Service quality—developing a service quality index in the provision of commercial bus contracts. Transp. Res. A Policy Pract. **37**(6), 499–517 (2003)
22. Kębłowski, W.: Free public transport: Scope and definitions. In: Dellheim, J., Prince, J. (eds.) Free Public Transit and Why We Don't Pay to Ride Elevators, pp. 1–6. Black Rose Book, Montreal (2018)
23. Fearnley, N.: Free fares policies: impact on public transport mode share and other transport policy goals. Int. J. Transp. **1**(1), 75–90 (2013)
24. Storchmann, K.: Externalities by automobiles and fare-free transit in germany—a paradigm shift? J. Public Transp. **4**(6), 89–105 (2003)
25. Baum, H.J.: Free public transport. J. Transp. Econom. Policy. **7**(1), 3–19 (1973)
26. De Witte, A., Macharis, C., Mairesse, O.: How persuasive is 'free' public transport?: a survey among commuters in the Brussels Capital Region. Transp. Policy. **15**(4), 216–224 (2008)
27. Canters, R.: Hasselt cancels free public transport after 16 years (Belgium). Eltis. (2014). http://www.eltis.org/discover/news/hasselt-cancels-free-public-transport-after-16-years-belgium-0
28. Van Goeverden, C., Rietveld, P., Koelemeijer, J., Peeters, P.: Subsidies in public transport. Eur. Transp. **32**, 381–404 (2006)
29. Boussauw, K., Vanoutrive, T.: Transport policy in Belgium: translating sustainability discourses into unsustainable outcomes. Transp. Policy. **53**, 11–19 (2017)
30. Cats, O., Susilo, Y.O., Remail, T.: The prospects of fare-free public transport: evidence from Tallinn. Transportation. **5**(40), 1083–1104 (2017)
31. Thøgersen, J.: Promoting public transport as a subscription service: effects of a free month travel card. Transp. Policy. **16**(6), 335–343 (2009)
32. Thøgersen, J., Møller, B.: Breaking car use habits: the effectiveness of a free one-month travel card. Transportation. **35**(3), 329–345 (2008)
33. De Witte, A., Macharis, C., Polain, C., Lannoy, P., Vandewalle, S., Steenberghen, T.: The impact of "free" public transport: the case of Brussels. Transp. Res. A Policy Pract. **40**(8), 671–689 (2006)
34. Steg, L.: Car use: lust and must. Instrumental, symbolic and affective motives for car use. Transp. Res. A Policy Pract. **39**(2-3), 147–162 (2005)
35. Tertoolen, G., van Kreveld, D., Verstraten, B.: Psychological resistance against attempts to reduce private car use. Transp. Res. A Policy Pract. **32**(3), 171–181 (1998)
36. Kos-Łabędowicz, J., Urbanek, A.: The potential for the development of carsharing and carpooling systems: a survey-based analysis of University students in Poland. Int. J. Transp. Econom. **44**(3), 399–425 (2017)
37. Stradling, S.G., Meadows, M.L., Beatty, S.: Identity and independence: two dimensions of driver autonomy. In: Grayson, G.B. (ed.) Behaviour Research in Road Safety. Transport Research Laboratory, Crowthorne, Berkshire (1999)

Methodology for the Evaluation of Walking Trips Among the Inhabitants in the Light of Marketing Research Results in Warsaw and Gdynia (Poland)

Katarzyna Hebel and Marcin Wołek

Abstract This chapter investigates walking mobility as an element of modal split on the example of cities of Warsaw and Gdynia (Poland). Research methodology presents a real challenge in trying to establish the precise share of walking in urban modal split. Whilst the methods applied in defining and measurements of mechanised trips have an established presence in the transport economics, the measurement of walking mobility is not as simple. The results vary not only because of the nature of cities (their size, spatial, demographic and socio-economic and natural features) but also because of methodology being used to measure walking trips. The chapter identifies main challenges of measurement of walking mobility, focusing on results of different research methods. The comparison was made using the case study of two cities in Poland conducted in 2015. Conclusions are of practical and methodological importance for future research of sustainable mobility policy.

Keywords Walking · Modal split · Urban mobility

1 Introduction

After the initial boom in the use of cars in urban travel, in the twenty-first century drawbacks associated with this mode of transportation have begun to emerge. Such drawbacks include congestion that significantly prolongs the duration of travel; air pollution; problems with parking and lack of space within cities; as well as health-related issues connected with over-reliance on car transport and the resulting lack of exercise [1–3]. Towards the end of the 1960s, some European cities have begun to

K. Hebel · M. Wołek (✉)
Faculty of Economics, University of Gdansk, Gdansk, Poland
e-mail: katarzyna.hebel@ug.edu.pl; marcin.wolek@ug.edu.pl

M. Suchanek (ed.), *Transport Development Challenges in the 21st Century*, Springer Proceedings in Business and Economics,
https://doi.org/10.1007/978-3-030-50010-8_20

plan the development of urban transportation systems which would take into account the needs of all road users, with special focus directed to pedestrian travel [4].

Nevertheless, the European Commission drew attention to the negative impact of transport on the natural environment within urban areas. Its claims were published in 1992 in the Green Book "Influence of transport on the environment. EU strategies for sustainable mobility" [5].

At the same time the sustainable development policy was introduced to the mainstream transport and mobility policies of many European cities. Within the policy preference is given to sustainable travel within cities, namely travel on foot, by bicycle, or by public transport. Many forms of combined travel are being promoted in order to decrease the importance of cars in everyday urban travel (i.e. walking and public transport, P&R, B&R), in addition to various forms of shared transportation (carpooling, carsharing) [6].

The growing importance of active mobility (walking and cycling) in urban areas represents one of the main current trends in sustainable mobility. The increase in the share of walking trips in the modal split within cities is perceived as one of the most effective economic, social [7], health-related [8], environmental and transport-related [9] solutions. An increase in mobility and accessibility together with a decrease in the role of cars and the resultant decrease of the demand for parking spaces [10] and the improved safety of all transport users, are named amongst the many transport benefits. Increasing share of active modes (walking and cycling) in school trips is an important area of decreasing car dependency in cities.

The share of any mode of transport within cities is varied and influenced by many factors such as demographics; structure of urban areas; stimulus and ways used to encourage the use a particular mode of travel; impact of a given mode of travel on the urban environment; accessibility of particular modes of travel in a given urban area.

Modal split is calculated on the basis of comprehensive research into mobility or transport behaviour of residents of particular areas [11]. It is mostly arrived at using the number of trips rather than transportation work (in passenger-kilometre), inclusive of distance travelled.

In recent years secondary data, available due to the increase in popularity of smartphones, has grown in importance. In such cases the following could serve as sources of information:

- Mobile telephones data from GSM operators
- GPS passive data [12]
- Bluetooth
- Wi-Fi
- Mobile apps

Research methodology and organisation present a real challenge in trying to establish the precise share of walking trips in urban modal split. Whilst the methods applied in arriving at definitions and measurements of motorised trips (including cycling) have an established presence in the transport economics, the measurement of pedestrian trip is not as simple. The following issues can be named as the main obstacles in this line of research: the incidentalness in its time and space aspects of pedestrian trip; difficulty to compare different areas of the city (which vary in the way they appeal to respondents, according to their density [13] and land use structure and diversity [14]; the extensive time required to gather data; important segments of citizens partially excluded from modern ways of measurements, including smartphones (i.e. kids and elderly people). Additionally, depending on the applied methods of research, the number of pedestrian trips on very short distances may be underestimated, as was the case during the process of collecting data as part of England's NTS, where respondents identified fewer short trips (below 50 yards) in their dairies as the week progressed. Revision of the methodology resulted in an increase of five short walking trips per person over a period of 12 months [15]. Moreover, the weather could have a powerful influence on the decision to travel on foot or in a different way [16, 17]. The sole definition of walking trip is ambiguous and can be understood in many ways, depending on the applied methods of research. It should be remembered that walking trip accompanies every other form of mechanised travel (by car, bicycle, public transport). The sustainable transport plan 2006–2016 by The Auckland Regional Transport Authority (ARTA) found that around 40% of short journeys (less than 2 km) were made entirely on foot and most trips included a walking component as some part of the journey [18]. Other researchers implemented a definition of "short trip" that lasts up to 10 min. From them a "proximity trip" was excluded, a trip made only by non-motorised modes [19]. Usually such trips cover short distances and despite the fact their share in the general number of trips may be high, their volume measured in passenger-kilometres will be much lower than in the case of mechanised means of urban mobility. The findings of the research conducted in 2017 as a part of England National Travel Survey seem to support this claim. Walking trips constituted 26% of the total number of trips but they covered only 3% of total distance made by all modes of transport [20] (see Table 1). Nevertheless, all the above-mentioned obstacles should not overshadow the benefits resulting from the increase of the share of walking in modal split [21].

The increase in significance of sustainable urban mobility make it indispensable to broaden research into walking, especially within city centres. The usefulness of such data surpasses merely its practicality in spatial and mobility planning. It can be potentially useful to the institutions of the public sector (i.e. health and wellbeing) and private companies (mostly for trade).

Table 1 Share of selected transport modes in number of trips and in distance

Modal split	Share in number of trips [%]	Share in distance [%]
Car	61	78
Train	2	8
Cycling	2	1
Walking	26	3

Source: based on National Travel Survey: England 2017, Statistical Release July 2018, https://assets.publishing.service.gov.uk/government/uploads/system/uploads/attachment_data/file/729521/national-travel-survey-2017.pdf

Taking particular action within the framework of sustainable development of transport requires first and foremost identification of the transport behaviour of residents. This article aims to present and compare the current methodology of research of walking mobility applied in two selected Polish cities.

The following research thesis was applied: the share of walking in urban modal split strongly depends on the methods being used to measure it.

Verification of such thesis needs to answer on specific research questions, namely:

- Which methods are basic for determining the share of walking trips in the modal split of Polish cities?
- Is the electronic (internet) survey a valuable source of information about walking mobility?
- What is the role of observation in determining the share of walking trips in modal split?

The thesis and research questions detailing them determined the structure of the paper. Case studies were used to answer the research questions. Two Polish cities were selected as part of it. The selection criterion was the fact of regularly conducting research on the mobility of residents (in Warsaw since 1980, in Gdynia since 1992) and the use of diverse research methods.

2 Results: Walking Trips in Warsaw

In some Polish cities there is a well proven methodology of complex traffic research. The first research of this type was carried out in Warsaw in 1980 [22]. Recent studies have been very comprehensive and included the study of transport behaviours of the residents of Warsaw, measurements of traffic and demand of public transport as well as examination of freight traffic in the agglomeration area [23].

It was established during the Warsaw Traffic Research of 2015 that 17.9% of trips can be classified as walking trips, which was defined as trips that are part of a trip completed via means of public transport (accessing stops and changing lines) [23]. Since Warsaw Traffic Research was conducted in 1980, 1993, 1998, 2005, and 2015, it is possible to compare data obtained over 35 years. A significant drop in the volume of the share of walking trips was noted, with its most dramatic decline taking place between the research of 1993 and 1998 (nearly 10%). During the 1980s and at the beginning of the 1990s nearly one-third of all trips were completed on foot. Currently walking trips have a share of less than one-fifth of all urban trips (Fig. 1).

The results of the Warsaw Traffic Research of 2015 were compared with the results obtained via internet survey conducted in the same year. Internet survey was also a part of the complex research (Table 2).

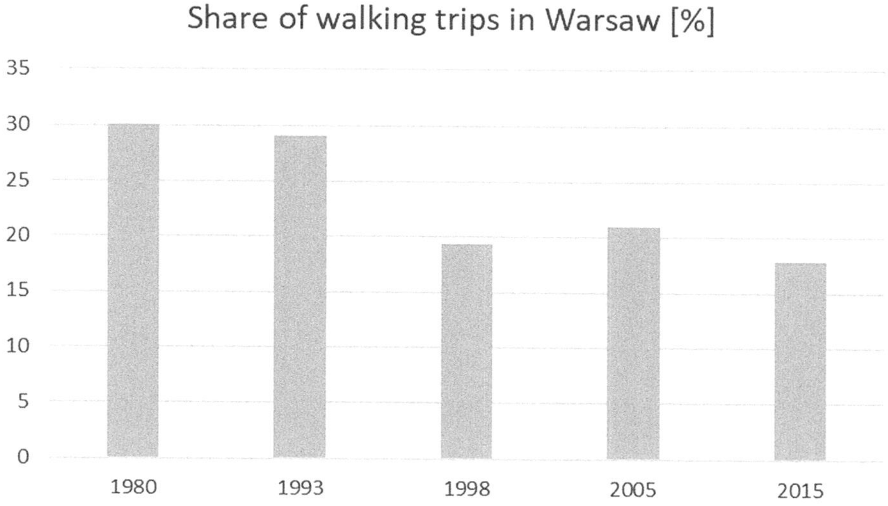

Fig. 1 Share of walking trips in Warsaw in the years 1980–2015 (%). Source: based on [23]

Table 2 Comparison of walking trips research of 2015 in Warsaw

Specification	Administrator	Area	Method	Sampling method	Sample
Main study of Warsaw Traffic Research	City of Warsaw	Warsaw	Personal, in-house interview	Random route	17,000
Internet survey	City of Warsaw	Warsaw and its functional area	CAWI	No sampling	6958

Source: [own research]

The results vary significantly (Fig. 2). According to the data obtained via internet questionnaire, the share of walking amounted only to 8.7% of all urban trips within Warsaw [24] and share of public transport trips dominated the whole results of the internet study. This goes to show the influence of methodology on the results and it particularly affects walking trips.

In addition, the results were split into age intervals of respondents involved in both research. The internet research strongly underestimated share of walking trips in all groups based on age but the biggest difference was visible for the youngest group of respondents (Fig. 3). Results of Complex Traffic Research support thesis of higher walking activity of youngest citizens, as they have less alternatives and lower level of independent travel decisions.

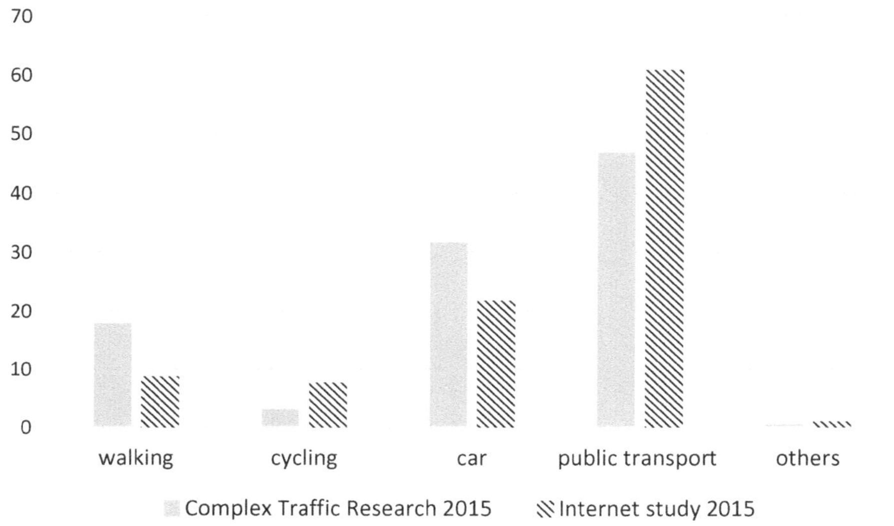

Fig. 2 Modal split in Warsaw in 2015 according to the method of research. Source: based on [23]

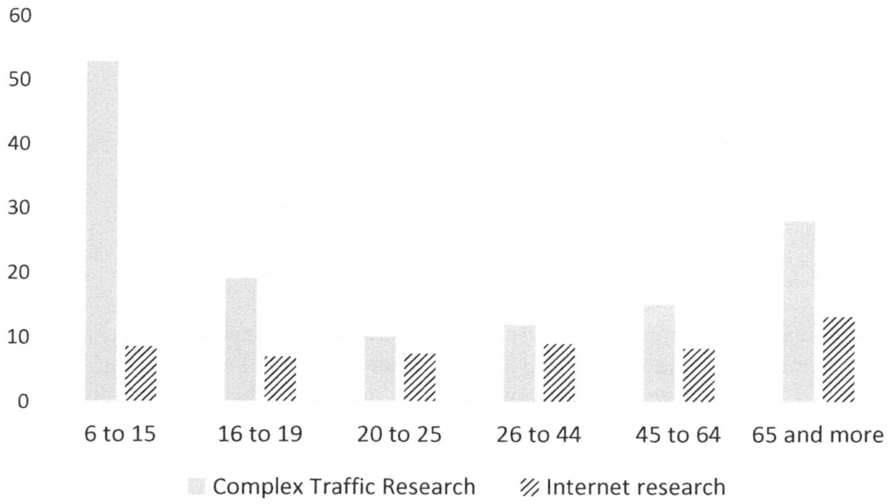

Fig. 3 Share of walking trips in modal split in Warsaw in 2015 according to the method of research and age interval of the respondents. Source: own study based on [24]

3 Results: Walking Trips Within Gdynia

In order to answer the research questions, the following part was dedicated to the analysis of the share of walking trips in research that was conducted within varied areas and according to diversified methods in city of Gdynia (Poland) (Table 3).

Table 3 Comparison of walking trips research of 2014–2016 in Gdynia

Specification	Administrator	Area	Method	Sampling method	Sample
Research of transport behaviour and preferences of the inhabitants of Gdynia	University of Gdansk and Public Transport Authority in Gdynia	Gdynia (all districts)	Personal, in-house interview	Random sample (based on the sex, age interval and district of living)	1% of inhabitants between 15 and 75 years of age (total size of sample: 2000 respondents)
Research of traffic structure and volume within Gdynia city centre	University of Gdansk and Roads and Greenery Authority in Gdynia	ul. Świętojańska (city centre of Gdynia)	Observation	n.a.	n.a.

Source: [own research]

Research into travel behaviour and preferences has been conducted in Gdynia since 1994, in the form of an in-depth structured interview. The sample consists of 1% of inhabitants between 15 and 75 years of age. The sample is chosen proportionally to the current structure of the city's population and analysed according to the cross-section of the number of inhabitants within a district, as well as their gender and age. The paper presents the results of the 2015 research.

The methodology used was based on a so-called "travel diary" where the respondents were asked to record all details of journeys completed on the previous day. For a trip to be classified as pedestrian, the distance covered on foot had to be longer than 500 m. Walking trips were, therefore, inclusive of trips solely covered on foot as well as trips were walking was just part of a longer journey, i.e. walking to a bus stop.

In this research walking trips amounted to 10.9% of all trips completed by the inhabitants of Gdynia in 2015.

In the search of determinants characteristic for walking trips, their nature was examined, including the age of inhabitants, their social and employment status as well as gender. Figure 4 presents the influence of the age of the respondents on the share of walking trips in the modal split for Gdynia. The lowest share of walking was specific to the group between 41 and 50 years (professionally active), whilst the highest was noted in the group of inhabitants between 61 to 70 years of age. Slightly higher (than the average) share of walking was also characteristic feature for respondents between 16 and 20 years. It has to be added, however, that the research did not include inhabitants under 16 years of age, for which walking is the most dominant way of urban mobility.

In the years 2014–2016 research into the structure and volume of traffic on Świętojańska Street in Gdynia (city centre) was conducted as a part of CIVITAS DYN@MO project. This article presents results from the 2015 research. That same

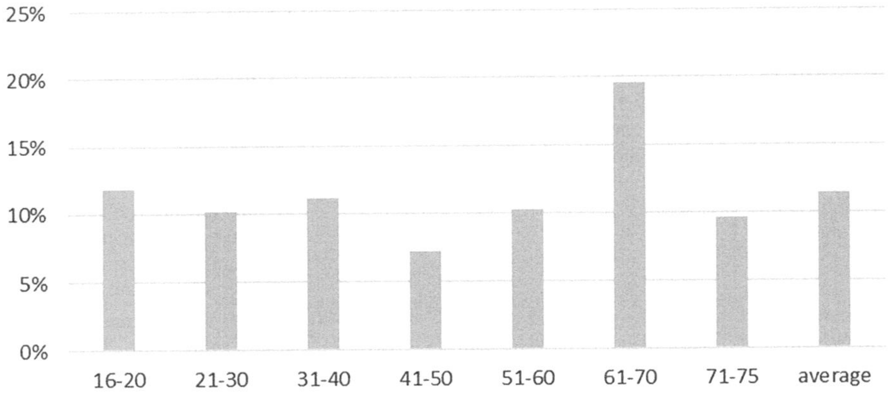

Fig. 4 The share of walking trips in the modal split for Gdynia in 2015 taking into account the age interval of respondents [in %]. Source: own work based on [25]

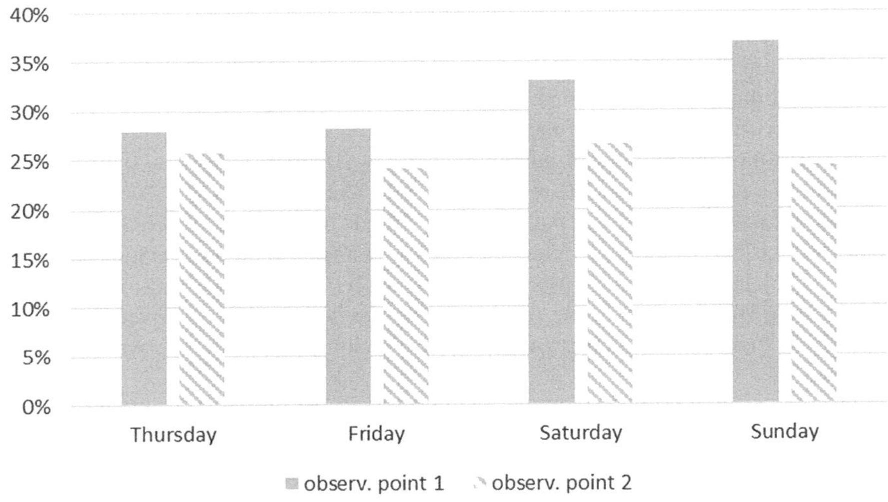

Fig. 5 The share of walking within Świętojańska Street in Gdynia in 2015 according to the day of the week and measurement point (hour of 5.00 am–10.00 pm). Source: own work based on [26]

year the previously mentioned research into travel behaviours and preferences was also conducted. For the purposes of this research the whole traffic including cars, cycling, public transport (trolleybuses) and pedestrians measured at 2 points on Świętojańska Street was taken into account (Fig. 5). Świętojańska Street is one of the most important streets in Gdynia. Apart from its historical and symbolic value, its location and functions it currently fulfils as an important element of the public transportation network within the city centre of Gdynia (13 out of 16 trolleybus lines pass via this street).

The adopted methodology of observation allowed traffic volume to be measured in both directions as well as for the division into basic means of travel within the centre of Gdynia. The results revealed a significant 28% share of walking within the city centre on a week-day, a result which is similar to other European cities, and as much as 37% on a Sunday. The results also differ according to observation point, especially for weekends. The results lead to the conclusion that share of pedestrian traffic is higher at weekends. This points to a strong correlation with the time inhabitants may devote to a trip on particular days of the week and also shows that the walking trips completed at weekends are mostly for pleasure—seen not as a necessary form of travel but a leisure activity.

Higher share of walking trips as a result of the observation might be explained due to the fact that there were no limitations in walking trip distance. Another factor is the attractiveness of the central part of the city and its "common" use, especially for leisure activities.

4 Discussion

Walking is becoming a frequent and strategic subject of scientific research and transport policy at local level [27]. The importance of walking goes far beyond mobility as it affects health topics as well [28]. A study conducted in Scotland has found that higher distance travelled by active modes of transport (mainly walking) characterises residents of the least deprived areas [29].

Although it is seen as an important tool to decrease greenhouse gas emissions [30, 31] its share in modal split in Polish capital city of Warsaw decreased substantially in period between 1980 and 2015.

Due to the nature of cities (their spatial structure, management, available public transport vehicles) results of analysed cities being used as case studies vary significantly. These differences are also a result of the application of different research methods.

There are no common and widely agreed criteria defining a walking trip (time, distance, or destination/motivation). Such a lack of unambiguous criteria diversifies the share of walking in the urban modal split [32]. The methodology applied by the National Travel Survey allows the shortest pedestrian trips, at a distance of up to 50 yards [20], to be singled out. In some complex traffic research in Poland a criterion of a minimum distance of 250 m was used [33].

Solid research findings could serve as the basis for the inclusion of walking trips into mobility structure. Such results should be characterised by regular, comprehensive and comparable comparisons in space and time. It is also necessary to broaden the research into walking trips through qualitative (matrix O-D, motivation, relations with built environment) and quantitative parameters (i.e. duration, distance, topography). Results of such research will allow for the inclusion of the needs of pedestrians in projects focused on transformation of street spaces.

Another issue is differentiation between walking trips that are independent and those which are part of a longer journey (as per accessing a bus stop, changing lines, or reaching a final destination after travelling on public transport or by car). Such differentiation requires a precise measuring tool. Such a tool can be a questionnaire precisely developed to highlight a complex nature of walking activity in a city.

5 Conclusions

In-house interview results are the basic method for determining walking in the modal split of both cities. The difference in the results is mainly due to a greater range of age limits for respondents in Gdynia (respondents aged 15–75) and recognition of travels over 500 m for walking.

It should be emphasised, however, that the methodology of research in Gdynia has been subordinated to public transport goals for years (the manager of research is the Public Transport Authority in Gdynia), and walking have been included in the research since 2015.

Research in the form of an internet survey should be considered as supplementary in the light of the results from Warsaw. At the same time, limitations of this method of contact with the respondent appeared (underestimation of young and old people, freedom in considering what could be treated as a walking trip).

An example of research dedicated to the identification of the nature and scope of walking trip can be seen in the research conducted in Gdynia within the main passageway of Świętojańska Street. The results confirmed a high share of pedestrian travel in the modal split for the city's centre. It is comparable with other European cities, amounting to 28% on a typical week-day. Moreover, the results allow for the conclusion that the share of walking trips rises at weekends. This points to a strong correlation with the time inhabitants may devote to a trip on particular days of the week and also shows that the walking trips completed at weekends are mostly for pleasure—seen not as a necessary form of travel but a leisure activity. Observation conducted in a comprehensive manner and covering all cross-sections of transport means is supplementary but a valuable method of verifying data obtained from other sources. It also allows to observe changes in the modal split during the whole week. It enables to highlight the differences between weekdays and days off, underlying the importance of leisure activities in urban mobility.

References

1. Aldred, R., Croft, J.: Evaluating active travel and health economic impacts of small streetscape schemes: an exploratory study in London. J. Transp. Health. **12**(2018), 86–96 (2019). https://doi.org/10.1016/j.jth.2018.11.009
2. European Commission: Action plan on urban mobility. COM(2009) 490 Final, 1–13 (2009)

3. Saunders, L.: Transport and Health in London. The main impacts of London road transport on health. (June). https://www.london.gov.uk/sites/default/files/transport_and_health_in_london_march_2014.pdf (2014)
4. City of Vienna: Urban mobility plan Vienna. https://www.wien.gv.at/stadtentwicklung/studien/pdf/b008443.pdf (2014)
5. Wolek, M., Wyszomirski, O. (eds.): The Trolleybus as an Urban Means of Transport in the Light of the Trolley Project. University of Gdansk Press, Sopot (2013)
6. Hebel, K., Wołek, M.: Change trends in the use of passenger cars on urban trips: car-pooling in Gdynia. Sci. J. Siles. Univ. Technol. Series Transp. **96**(96), 37–47 (2017). https://doi.org/10.20858/sjsutst.2017.96.4
7. Wang, Y., Chau, C.K., Ng, W.Y., Leung, T.M.: A review on the effects of physical built environment attributes on enhancing walking and cycling activity levels within residential neighborhoods. Cities. **50**, 1–15 (2016). https://doi.org/10.1016/j.cities.2015.08.004
8. Arranz-López, A., Soria-Lara, J.A., López-Escolano, C., Pueyo Campos, Á.: Retail mobility environments: a methodological framework for integrating retail activity and non-motorised accessibility in Zaragoza, Spain. J. Transp. Geograp. **58**, 92–103 (2017). https://doi.org/10.1016/j.jtrangeo.2016.11.010
9. Özdemir, D., Selçuk, İ.: From pedestrianisation to commercial gentrification: the case of Kad ı köy in Istanbul. Cities. **65**, 10–23 (2017)
10. Soni, N., Soni, N.: Benefits of pedestrianization and warrants to pedestrianize an area. Land Use Policy. **57**(5), 139–150 (2016). https://doi.org/10.1016/j.landusepol.2016.05.009
11. Spencer, A.H., Andong, W.: Light rail or busway? A comparative evaluation for a corridor in Beijing. J. Transp. Geograp. **4**(4), 239–251 (1996). https://doi.org/10.1016/S0966-6923(96)00027-0
12. Lue, G., Miller, E.J.: Estimating a Toronto pedestrian route choice model using smartphone GPS data. Travel Behav. Soc. (2019). https://doi.org/10.1016/j.tbs.2018.09.008
13. Ton, D., Duives, D.C., Cats, O., Hoogendoorn-Lanser, S., Hoogendoorn, S.P.: Cycling or walking? Determinants of mode choice in the Netherlands. Transp. Res. A Policy Pract. **123**(August 2018), 7–23 (2019). https://doi.org/10.1016/j.tra.2018.08.023
14. Ewing, R., Cervero, R.: Travel and the built environment. J. Am. Plan. Assoc. **76**(3), 265–294 (2010). https://doi.org/10.1080/01944361003766766
15. Office for National Statistics: National Travel Survey Quality Review. (January), 1–17. https://assets.publishing.service.gov.uk/government/uploads/system/uploads/attachment_data/file/775062/annex-d-nts-2019-quality-report.pdf (2019)
16. Caprì, S., Ignaccolo, M., Inturri, G., Pira, M.L.: Green walking networks for climate change adaptation. Transp. Res. D Transp. Environ. **45**, 84–95 (2016)
17. Vanky, A.P., Verma, S.K., Courtney, T.K., Santi, P., Ratti, C.: Effect of weather on pedestrian trip count and duration: city-scale evaluations using mobile phone application data. Prevent. Med. Rep. **8**, 30–37 (2017)
18. Auckland Regional Transport Authority: Sustainable transport plan 2006-16. https://at.govt.nz/media/imported/4827/AT_ARTA_Policy_SustainableTransportPlan2006_16.pdf (2006)
19. Marquet, O., Miralles-Guasch, C.: The Walkable city and the importance of the proximity environments for Barcelona's everyday mobility. Cities. **42**(PB), 258–266 (2015). https://doi.org/10.1016/j.cities.2014.10.012
20. National Travel Survey: 2017. National Travel Survey: England. Statistical Release, (July). https://assets.publishing.service.gov.uk/government/uploads/system/uploads/attachment_data/file/729521/national-travel-survey-2017.pdf (2018)
21. Wolek, M.: The significance of pedestrian mobility as shown by the example of the city of Gdynia. In: Bąk, M. (ed.) Transport Development Challenges in the Twenty-First Century: Proceedings of the 2015 TranSopot Conference. Springer, Cham (2016)
22. Chitulescu, Z., Gołaszewska, G., Hoch, J., Kawecka-Monkiewicz, U., Ledzion, G., Pieńkowski, B., Ujazdowski, Z.: Warszawskie Badanie Ruchu 1980.pdf. http://transport.um.warszawa.pl/sites/default/files/KBR 1980 I Raport_1.pdf (1984)

23. Kostelecka, A., & Kulpa, T. The Warsaw Traffic Study in 2015 along with the development of a traffic model. Synthesis. http://transport.um.warszawa.pl/sites/default/files/WBR 2015 SYNTHESIS EN.pdf (2016)
24. Kostelecka, A.: Analiza wyników badania internetowego [Analysis of internet research results. Warsaw Traffic Research 2015]. http://transport.um.warszawa.pl/warszawskie-badanie-ruchu-2015/wyniki-wbr-2015 (2016)
25. Gdynia, Z.K.M.: Preferencje i zachowania komunikacyjne mieszkańców Gdyni. Raport z badań marketingowych 2015. [Preferences and transport behavior of citizens of Gdynia. Marketing research report 2015]. https://zkmgdynia.pl/files/Pliki%20do%20pobrania%20-%20inne/A4_zkmgdynia-PZKMG-raport%202015.pdf (2016)
26. Wołek, M.: Badania natężenia i struktury ruchu na ul. Świętojańskiej w Gdyni w 2015 roku [Research on traffic volume and structure on the Świętojańska Street in Gdynia in 2015]. CIVITAS DYN@MO Project Report (2015)
27. Tight, M., Timms, P., Banister, D., Bowmaker, J., Copas, J., Day, A., et al.: Visions for a walking and cycling focussed urban transport system. J. Transp. Geograp. **19**(6), 1580–1589 (2011). https://doi.org/10.1016/j.jtrangeo.2011.03.011
28. Singleton, P.A.: Walking (and cycling) to well-being: modal and other determinants of subjective well-being during the commute. Travel Behav. Soc. **16**, 249–261 (2018). https://doi.org/10.1016/j.tbs.2018.02.005
29. Olsen, J.R., Mitchell, R., Mutrie, N., Foley, L., Ogilvie, D.: Population levels & inequalities in active travel Scotland 2017. Prevent. Med. Rep. **8**, 129–134 (2017)
30. European Commission: Sustainable urban mobility: European policy, practice and solutions. (2017) https://doi.org/10.2832/51274
31. European Union, United Nations Human Settlements Programme. The State of European Cities **2016** (2016). https://doi.org/10.2776/636682
32. European Platform on Mobility Management: TEMS—The EPOMM Modal Split Tool. http://www.epomm.eu/tems/upload.phtml (n.d.). Accessed 19 Nov 2019.
33. Vistula, V., Gdańska, B.R.: Raport - Gdańskie Badania Ruchu 2016. http://www.brg.gda.pl/attachments/article/282/Raport-III.pdf (2016)

Changes in Urban Transport Behaviour and Preferences of Residents in Employment: Gdynia Case

Krzysztof Grzelec and Justyna Staszak-Winkler

Abstract Urban transport policies are currently determined by sustainable mobility. The selection and application of appropriate tools, which takes into account transport behaviour and preferences of residents, as well as the analysis of their willingness to change those, are necessary in order to achieve the goals of sustainable mobility. The identification of these determinants and trends requires research to be carried out among the entire cross-section of residents, as well as on particular groups of consumers targeted by the services offered. Working people constitute one of the basic segments of consumers, regardless of the way in which urban trips are completed. The article presents an analysis of changes in transport preferences and behaviour of the aforementioned group of employed consumers resident in Gdynia. It compares results of comprehensive research into transport behaviour and preferences carried out in 2008 and 2015 and shows to what extent over a 7-year period those preferences and behaviour of people in employment changed compared to the entire cross-section of residents between 15 and 75 years of age. During this period an increase in the number of cars per household (from 64% to 72%) as well as an increase in their share in urban trips (from 47% to 58%) was noted. It also evaluates the effectiveness of measures applied in order to achieve more sustainable transport.

Keywords Sustainable mobility · Transport preferences · Transport behaviour · Working residents · Residents in employment

K. Grzelec (✉)
Faculty of Economics, University of Gdansk, Gdansk, Poland
e-mail: krzysztof.grzelec@ug.edu.pl

J. Staszak-Winkler
Faculty of Civil and Environmental Engineering, Gdansk University of Technology, Gdansk, Poland

M. Suchanek (ed.), *Transport Development Challenges in the 21st Century*, Springer Proceedings in Business and Economics,
https://doi.org/10.1007/978-3-030-50010-8_21

1 Introduction

The research and analysis of transport behaviour and preferences are indispensable in the effective implementation of the tools of sustainable mobility. The term *change in transport behaviour* is vague and refers to both the observed trends as well as the intentions of entities providing services. Transport behaviour is changing, and understanding these changes can, in turn, provide the opportunity to further influence that behaviour. Politicians and those responsible for transport organisation have greater influence on transport behaviour than they realise. However, that influence exists only under the provision that transport services are cohesive, sustained over a long period of time, and supported by analytical data and evaluation [1]. It has been observed that educational and information tools may have a substantial impact on transport behaviour, however, they need to be evaluated with regards to their effectiveness for the most viable techniques, procedures, and target groups to be identified [2]. Research in the Netherlands suggests that some changes in transport behaviour can be supported by monetary incentives. It has been proven that employers "monetary incentives change employees" willingness to avoid rush hour travel [3]. Research into changes in the transport behaviour of people in employment, encouraged by certain incentives (i.e. monetary) were also presented by Karney [4]. Significant issues with air quality in southern California impelled the South Coast Air Quality Management District to issue a directive requiring employers to draft and implement programmes to reduce employee commuting times [5]. A review of changes in tax law to favour more environmentally friendly forms of travel and, in particular, to encourage employers to take part in Transport Demand Management programs in Great Britain was presented [6].

For over 30 years, there was a visible tendency of societies of Central and Eastern Europe to motorise and increase the usage of cars. It has lead to an increase in undesired consequences, such as pollution, accidents and congestion. Changing this mind-frame, together with the corresponding transport behaviour, will require public transport to become a strong competitor against the perceived advantages of cars. Timetables, frequency of transport, comfort and safety of travel are one of the most important factors determining transport behaviour [7].

Transport behaviour is determined by the location of a person's place of work, whilst the choice of the workplace itself is influenced by the availability of transport options. Mixed land use, residential density, street connectivity, and commute distance have also been identified as potential variables affecting transport-related physical activity (TPA) behaviours [8]. Flexible working arrangements constitute a useful tool in influencing transport behaviours. Such arrangements may have a positive impact encouraging the choice of sustainable transport, and, in case where trips to a workplace are completed by car, allow for savings in both time and fuel [9].

Significant changes in urban travel market have already taken place in the twenty-first century with car-sharing and bike-sharing possibly pushing the shift in transport behaviours even further. Over the last 20 years car-sharing rose to the position of the main travel choice of over one million consumers worldwide. Car-sharing companies are currently in operation in over 1100 cities in 26 countries and on 5 continents [10]. It is anticipated that this form of travel will play its part in the reduction of the CO_2 emissions (especially due to the vapour injection technology in electric vehicles) and overall decrease in the number of cars within cities [11]. Research in Germany proved that car-sharing constitutes a suitable supplement to public transport [12]. Bike-sharing constitutes an interesting alternative in work commute. The wide number of benefits connected with this form of travel include: improvement in overall health; wider choice of transport options; more direct connections; possibility to decrease the duration of travel and costs connected with travel; ability to gain different experience to that of other modes of travel. These benefits are unevenly spread since consumers who opt for this type of travel are usually young men from social and economic upper classes. There is also insufficient evidence to prove that bike-sharing plays a significant part in decreasing the congestion, CO_2 emissions and pollution [13]. Further research is necessary. A steady growth in popularity of electric bike-sharing has been observed. The effectiveness of this mode of travel, however, requires both vehicles to be suitably adjusted technically as well as logistic adjustments to the network [14].

The increasing popularity of a healthy lifestyle encourages residents to travel short distances on foot. The share of pedestrians in travel depends on many factors. A British Study suggests that the increase in pedestrian travel is influenced by such factors as: pavement cleanliness, safe crossing places, good connectivity and sense of security [15]. Furthermore, an American study shows that the destinations of pedestrians mostly include: home, shopping, restaurants, bars, banks and other services. Distance travelled on foot to main destinations does not go above 600 m and on rare occasions exceeds 1200 m. It follows that the assumption made in literature on the subject of pedestrian travel, claiming that the distance to a pedestrian's main destination should not exceed 1000 m from their household, is entirely justified [16].

This chapter uses the results of comprehensive studies on transport preferences and behaviours of inhabitants in Gdynia, carried out in 2008 and 2015, analyzing changes in transport behaviours and preferences in the working people segment. It has been hypothesized that the city authorities' activities in achieving sustainable mobility goals are not delivering the expected results. The city's policy does not lead to the expected changes in modal split—reducing the share of travel by car. The reasons for choosing specific means of transport were also analyzed, examining their statistical significance in connection with modal split. The conclusions of the article should be used by city authorities when choosing specific instruments of sustainable mobility.

2 Methodology

The analysis of the transport behaviour of residents in employment was carried out on the basis of results obtained in the comprehensive research into transport needs, demand, transport behaviour and preferences among the residents of Gdynia.

The research was carried out in 2008 and 2015. Since 1994, every 2–3 years, the Public Transport Authority (PTA) has carried out research connected with transport behaviours of its consumers. The analysis of results obtained during such research is then prepared by academic workers experienced in both organisations of public transport services as well as designing transport systems.

The research was based on a standardised direct interview method. The measuring tool was made up of a questionnaire of 40–50 closed and open questions. The questionnaire was completed with respondents chosen on the basis of random sampling, who were resident in Gdynia and were between 15 and 75 years of age. The top and bottom age cap was established on the basis of mobility criteria, independence in decision-making and rapport building ability. The sample consisted of 1% of randomly chosen residents proportionally to the population of every district. During the research interviewers had their primary and backup lists of respondents in case the first choice respondent was unable to respond to the survey. The reasons that prevented an interview with a respondent from being conducted were as follows: the owner of the property was letting his property or the person was too active professionally and therefore most likely to travel by car. Substituting the initial respondent usually was connected with different transport behaviour of the latter. Therefore, car share in modal split has to be assumed to be lower than in reality.

The database was processed in such a way as to obtain information on the transport behaviour and preferences of respondents in employment. A comparison analysis was applied with the support of statistical analysis. Participant observation was implemented to evaluate sustainable mobility policy in Gdynia.

3 Results

3.1 *Profile of the Segment Comprising of Respondents in Employment*

Table 1 presents the profile of the segment comprising of respondents in employment who were resident in Gdynia in the years 2008 and 2015.

The share of respondents in employment increased by 9% between the years 2008 and 2015. The participation of women in employment decreased marginally. The number of women employed in Poland has been decreasing since the launch of the governmental 500+ benefit scheme in 2017. The decrease in the number of women in employment in the data analysed may be connected with the aforementioned methodology and difficulty in reaching women who are highly active professionally. The share of people in employment who were under or 40 years of age decreased, whilst the number of respondents in employment who were between 61 and 70 years of age increased by 5%. This was connected with the changes introduced by the government at the time, which extended the age of retirement by 2 years for men (from 65 to 67) and by 5 years for women (from 60 to 65).

Table 1 Profile of the segment comprising of respondents in employment resident in Gdynia in the year 2008 and 2015 (percentage)

Profile	2008	2015
Share of respondents in employment	46.8	55.8
Women	48.4	45.7
Men	51.6	54.3
Age		
16–20	0.2	0.3
21–30	19.4	14.5
31–40	33.5	32.0
41–50	25.0	24.2
51–60	20.3	22.4
61–70	1.6	6.6
71–75	0.0	0.0
Car in the household	75.0	81.8
Main car user	81.5	80.7
Commuting to workplace	88.1	88.1

Source: own study based on the results of the PTA research

Table 2 Modal split for respondents in employment resident in Gdynia in the years 2008 and 2015

Modal split	2008	2015
Car	60.2	61.9
Public transport	38.8	36.3
Bicycle	0.7	1.3
Other	0.3	0.5

Source: own study based on the results of the PTA research

The number of households with a car notably increased by 6.8%. The share of main car users within a household and travelling to work did not change significantly within the analysed segment of respondents in employment.

3.2 *Transport Behaviour and Preferences of Respondents in Employment*

Table 2 presents the modal split for respondents in employment resident in Gdynia.

No significant changes in the modal split for respondents in employment resident in Gdynia can be seen during the analysed 7-year period. The share of trips completed by car increased by 1.7% and that of bicycles by 0.6%. At the same time the share of trips completed by public transport decreased by 2.5%. It can, therefore, be assumed that the segment of respondents in employment, despite a notable increase in the number of cars present in households, is characterised by the certain stability of transport behaviours (the research was not inclusive of trips completed on foot at a distance below or equal to 0.5 km).

Table 3 presents a comparison of modal split for all respondents resident in Gdynia. The transport behaviours of Gdynia inhabitants' have changed significantly in the years analysed. The share of trips completed by car in the years between 2008 and 2015 increased by 10.9% whilst the share of trips completed via public transport decreased by 12.2%. This proves that the change in the transport behaviours of respondents in employment was indeed insignificant. The comparison of modal split between Tables 2 and 3 indicates that significant changes in transport behaviour of the general cross-section of respondents were triggered by other segments of inhabitants (unemployed, students and pensioners) and the increase in the number of those respondents in employment.

Taking into account the goals of sustainable mobility policy the aforementioned trend has to be perceived as unfavourable since the increase in the number of respondents in employment among which the number of car trips was the highest has been coupled with the pro-car modal shift in other segments of respondents. As a result, the number of cars present on the roads is increasing not only during the peak hours but also during other times of the day.

It has to be added that this trend applies only to trips completed by privately owned cars as car-sharing scheme was launched in Gdynia in 2017.

Due to the low volume of bicycles participating in trips in 2008 this mode of transportation was not included in the close research of transport behaviours at that time. The increase in their volume over the following years impelled the PTA to include bicycles in their research in 2015. Table 4 presents the basic data regarding the ways in which respondents in employment use bicycles during their commute. The most prevalent among the reasons influencing the decision to commute to work by bicycle were: pleasure and health reasons.

Table 5 presents the reasons behind choosing to make urban trips by car.

The reasons why respondents in employment decide to make urban trips by car did not change significantly. It has to be stressed that a different scale was applied in the research of 2015 as compared to the research in 2008 (more detailed reasons). For example, no need to wait specified in 2008 by 14% of respondents corresponds with two reasons specified in the research of 2015, namely: no need to wait and a direct connection, which were given by 17% of respondents. Despite difficulties with a conclusive comparison of findings between the research of 2008 and 2015, it

Table 3 Modal split for all respondents resident in Gdynia in the years 2008 and 2015

Modal split	2008	2015
Car	47.0	57.9
Public transport	52.2	40.0
Bicycle	0.4	1.8
Other	0.4	0.3

Source: own study based on the results of the PTA research

Table 4 Transport behaviour of respondents in employment commuting by bike (percentage)

Detailed list	2015
Share of households with a bicycle	63.9
Reasons for using a bicycle:	
Pleasure	52.9
Health reasons	26.8
Being eco-friendly	6.0
Cutting on travel costs	5.0
Shorter duration of travel	4.3
Enabling direct trip	4.3
Other	0.7

Source: own study based on the results of the PTA research

Table 5 Reasons of respondents in employment, resident in Gdynia in the years 2008 and 2015, for choosing to commute by car (percentage)

Reason	2008	2015
Safety—no threat of aggressive behaviour of other passengers	3.7	2.1
No need to travel to and from a bus stop	11.4	6.8
No need to wait	14.2	9.5
Direct connection	–	7.3
No tiresome company of other passengers	–	2.9
Shorter duration of travel	28.2	22.8
Lower costs	2.1	3.3
Travelling with kids	–	4.7
Transporting goods or groceries	–	9.4
Better comfort	28.5	22.3
Car as a work tool	8.8	6.8
Poor health	–	0.9
Other	3.1	1.2

Source: own study based on the results of the PTA research

can be stated that the main reasons behind the decision to travel by car, made by respondents in employment, are better comfort (inclusive of the flexibility that travel by car provides) and the shorter duration of travel. A comparison of the duration of work commute by car shows that an average duration of work commute by this mode of transport was shorter in comparison to work commute via public transport by 52% in 2008 and 51% in 2015.

Table 6 presents the reasons for choosing public transport for urban trips given by respondents who owned a car.

It has to be stressed that when analysing results a different scale was applied in both studies. The research of 2008 did not include parking fees among the reasons behind the decision of respondents in employment who owned a car to travel via public transport since the fees were introduced in 2008. The research of 2015

Table 6 Reasons for travelling via public transport on urban trips by respondents in employment, resident in Gdynia in the years 2008 and 2015 (percentage)

Reason	2008	2015
Car used by someone else	22.2	16.4
Unfit to drive	8.5	7.8
Lower cost of travel	13.8	11.9
Parking fees	–	13.0
Trouble with parking near destination	16.9	14.8
Weather	–	5.4
Good quality of public transport	8.4	10.3
Congestion	11.4	11.5
Technical issues with the car	5.9	6.2
Other	12.9	2.7

Source: own study based on the results of the PTA research

Table 7 Most important features of public transport defined by respondents in employment, resident in Gdynia in the years 2008 and 2015 (percentage)

Feature	2008	2015
Direct connection	18.7	17.9
Frequency	17.4	17.4
Accessibility	13.7	12.5
Reliability	5.5	8.2
Low cost of travel	6.6	8.1
Speed	8.3	7.0
Punctuality	18.0	17.8
Regularity	2.8	2.8
Exhaustive information	0.6	0.9
Comfort	5.5	4.1
Other	2.9	3.3

Source: own study based on the results of the PTA research

specified weather as one of the reasons to opt for travel via public transport (it was frequently mentioned in the research of 2008 and classified as an "other" response). The inaccessibility to the car was less frequently given as a reason to travel via public transport. This is understandable since the number of cars per household in Gdynia increased from 1.11 in 2008 to 1.21 in 2015.

Table 7 specifies the main features of public transport as defined by respondents.

The three most important features of public transport, as perceived by the respondents, in order of importance, were: direct connection, punctuality and frequency. Findings confirm earlier conclusions that the attractiveness of public transport for respondents in employment is determined mostly by factors connected with the duration of travel. It is, therefore, necessary to implement solutions that would decrease the disproportion in duration of travel by car and via public transport. Such solutions can be seen in bus passes and technologies which favour public transport on crossings with traffic lights. In 2017 Gdynia had a total of 3 km of bus passes, whilst Warsaw 44 km and other EU cities from 100 km to as much as 250 km.

The notion of "comfort" is very broad and may be understood as the availability of a seat, conditions of travel (comfortable temperature, space for luggage at no additional charge, availability of transport at a required time, etc.). Table 8 presents the comparison of comfort understood as the availability of a seat on board public transport vehicles.

The significance of comfort did not largely change in the period analysed (Table 8), however, the expectation to find an available seat always rose dramatically on 100% or almost a 100% of occasions (by 8–10%). This increase is specific to other segments of respondents and may be partially explained by the increase of respondents in employment who were between 61 and 70 years of age (by 5%). The reasons for the shift may also be seen in the changing conditions of urban travel in Gdynia between the years of 2008 and 2015. In the period analysed the offer for transport services expressed in vehicle-kilometres diminished by 2.5%. However, thanks to the smart management of services, the experienced quality did not suffer a slope since the rating of public transport organised and managed by the PTA, on a scale of 2 to 5 (2 being the lowest and 5 the highest), increased from 4.13 to 4.24 (within the segment of respondents in employment from 4.12 to 4.20). At the same time, the share of public transport in modal split decreased by 12%. Passenger volume, according to data gathered by the PTA, decreased between the years of 2008 and 2015 by 16%. This translated to an increase in the possibility to find a seat on board of public transport vehicles which, in turn, caused a shift in satisfaction levels regarding the expected comfort of travel. The improvement of travel conditions, although not anticipated by some passengers and, thus, bearing no influence on levels of their satisfaction, was accepted and became the expected standard of services provided. This conclusion is supported by the study of the factors which least lived up to expectations of passengers. Comfort of the second least frequently met factor of travel via public transport in 2008, rose to the fifth position in 2015, i.e. it was a more common factor in 2015 than in 2008. This proves that in addition to analysing the ratings of factors the occurrence of any qualitative changes must also be verified.

Table 8 Expected standard of travel as defined by respondents employed and resident in Gdynia in the years 2008 and 2015 (percentage)

Level of comfort	2008	2015
A seat always available	10.0	20.1
A seat mostly available	34.3	42.4
No seat but relatively comfortable	30.2	25.5
Acceptable when crowded	10.4	3.4
No opinion	15.1	8.6

Source: own study based on the results of the PTA research

4 Discussion and Conclusion

The analysis of transport behaviours of the respondents in employment provides new areas for discussion.

1. The results confirm that the cross-section of the respondents should be segmented according to differences typical for particular segments. It is purposeful not only to define segments but also to analyse changes in transport behaviours and preferences over time.
2. The increase in motorization within household members, coupled with the prevailing tendency of respondents in employment to commute by car, does not influence changes in behaviours. However, this segment may have a negative influence on the general modal split for respondents should its volume increase.
3. Lack of significant changes in features does not mean that their internal structure remains unchanged. The results suggest that further research into the internal changes of the said features, which may be defined as complex, should be carried out. Such features include:

    (a) Direct connection.
    (b) Accessibility.
    (c) Reliability.
    (d) Access to information.
    (e) Comfort.

4. The period chosen for the analysis of changes in transport behaviour and preferences of respondents in employment, resident in Gdynia, did not include either car-sharing (launched in 2017) or bike-sharing (launched in 2018). Certain changes in modal split have to, therefore, be anticipated as discussed by Dowling and Kent [10], Firnkorn and Muller [11], and Huwer [12]. The question of whether the new forms of travel will bear influence on changes among employed consumers, which are characterised by certain stability in their choices, remains open. Taking into account the modest share of bicycles in trips completed by employed respondents it can be concluded that Gdynia faces a seriously challenging task in the popularisation of bicycle travel, including the development of bicycle paths [17], improvement of cyclist safety [18], and encouragement for the employees to participate in cycle to work Schemes [19].

References

1. Goodwin, P., et al.: Changing Travel Behaviour. ESRC Transport Studies Unit, University College London. Preprint based on the final TSU conference, subject to revision. (2004)
2. Gärling, T., Fujii, S.: Travel behavior modification: theories, methods, and programs. In: Kitamura, R., Yoshi, T., Yamamoto, T. (eds.) The expanding sphere of travel behaviour research, pp. 97–128. Emerald, Bingley (2009)

3. Ben-Elia, E., Ettema, D.: Changing commuters' behavior using rewards: a study of rush-hour avoidance. Transp. Res. F Traff. Psychol. Behav. **14**(5), 354–368 (2011). https://doi.org/10.1016/j.trf.2011.04.003
4. Kearney, A.R., De Young, R.: Changing commuter travel behavior: employer-initiated strategies. J. Environ. Syst. **24**(4), 373–393 (2005). https://doi.org/10.2190/l2np-8aqm-fjrg-grpv
5. Giuliano, G., Hwang, K., Wachs, M.: Employee trip reduction in Southern California: first year results. Transp. Res. A. **27**(2), 125–137 (1993). https://doi.org/10.1016/0965-8564(93)90068-V
6. Nelson, N.: 7 best practices in transportation demand management. Transportation. **3**(Transportation Demand Management), 28 (2008). http://www.seattle.gov/transportation/docs/ump/07. Seattle Best Practices in Transportation Demand Management.pdf
7. Grdzelishvili, I., Sathre, R.: Understanding the urban travel attitudes and behavior of Tbilisi residents. Transp. Policy. **18**(1), 38–45 (2011). https://doi.org/10.1016/j.tranpol.2010.05.006
8. Badland, H.M., Schofield, G.M., Garrett, N.: Travel behavior and objectively measured urban design variables: associations for adults traveling to work. Health Place. **14**(1), 85–95 (2008). https://doi.org/10.1016/j.healthplace.2007.05.002
9. Ott, M., Slavin, H., Ward, D.: Behavioral impacts of flexible working hours. Transp. Res. Rec. **767**, 1–6 (1980)
10. Dowling, R., Kent, J.: Practice and public-private partnerships in sustainable transport governance: the case of car sharing in Sydney, Australia. Transp. Policy. **40**, 58–64 (2015). https://doi.org/10.1016/j.tranpol.2015.02.007
11. Firnkorn, J., Müller, M.: What will be the environmental effects of new free-floating car-sharing systems? The case of car2go in Ulm. Ecol. Econ. **70**(8), 1519–1528 (2011). https://doi.org/10.1016/j.ecolecon.2011.03.014
12. Huwer, U.: Public transport and car-sharing—benefits and effects of combined services. Transp. Policy. **11**(1), 77–87 (2004). https://doi.org/10.1016/j.tranpol.2003.08.002
13. Ricci, M.: Bike sharing: a review of evidence on impacts and processes of implementation and operation. Res. Transp. Bus. Manag. **15**, 28–38 (2015). https://doi.org/10.1016/j.rtbm.2015.03.003
14. Ji, S., Cherry, C.R., Han, L.D., Jordan, D.A.: Electric bike sharing: simulation of user demand and system availability. J. Clean. Prod. **85**, 250–257 (2014). https://doi.org/10.1016/j.jclepro.2013.09.024
15. Kelly, C.E., Tight, M.R., Hodgson, F.C., Page, M.W.: A comparison of three methods for assessing the walkability of the pedestrian environment. J. Transp. Geogr. **19**(6), 1500–1508 (2011). https://doi.org/10.1016/j.jtrangeo.2010.08.001
16. Millward, H., Spinney, J., Scott, D.: Active-transport walking behavior: destinations, durations, distances. J. Transp. Geogr. **28**, 101–110 (2013). https://doi.org/10.1016/j.jtrangeo.2012.11.012
17. Tilahun, N.Y., Levinson, D.M., Krize, K.J.: Trails, lanes, or traffic: valuing bicycle facilities with an adaptive stated preference survey. Transp. Res. A. **41**, 287–301 (2007)
18. Winters, M., Teschke, K.: Route preferences among adults in the near market for bicycling: findings of the cycling in cities study. Am. J. Health Promot. **25**(1), 40–47 (2010)
19. Hunt, J., Abraham, J.: Influences on bicycle use. Transportation. **34**(4), 453–470 (2007)

The Role of Urban Environmental Sound in the Orientation of People with Impaired Vision

Małgorzata Orczyk and Franciszek Tomaszewski

Abstract This chapter presents results of surveys conducted with blind people and analysed sound signals necessary in the development of the method of teaching spatial orientation to people with impaired vision. The survey was conducted to identify the most important problems faced by people with impaired vision living in a large urban agglomeration and to determine the impact, type and nature of sounds generated in the city in terms of spatial orientation. On this basis, more than 1700 sound situations of the city that are characteristic of a large urban agglomeration were selected for registering. The recorded sounds from characteristic places, vehicles and sound events were analysed in order to create a library of sounds that will be used to teach spatial orientation.

Keywords Sounds of the city · People with impaired vision · Spatial orientation

1 Introduction

For people with impaired vision, the restriction of free movement is the most obvious 'barrier' of life. This leads to loss of self-esteem, makes it difficult to get different kinds of experiences, makes one dependent on a guide and on the surroundings. Despite great progress of information technology and development of hardware technologies, teaching people with impaired vision how to move in the near and distant areas uses old, not always effective methods. In teaching spatial orientation, the most important sense for people with impaired vision, which is hearing, is not sufficiently used. Currently, more and more specialised electronic equipment is appearing on the market, which is designed to assist people with impaired vision in spatial orientation. Not all of these devices have been accepted by the blind community, and not only because of their very high price, but also because of simple

M. Orczyk (✉) · F. Tomaszewski
Faculty of Transport Engineering, Poznan University of Technology, Poznan, Poland
e-mail: Malgorzata.Orczyk@put.poznan.pl

M. Suchanek (ed.), *Transport Development Challenges in the 21st Century*, Springer Proceedings in Business and Economics,
https://doi.org/10.1007/978-3-030-50010-8_22

resistance to new technologies and technical novelties. Therefore, it is advisable to use in the spatial orientation of people with impaired vision a natural method based on hearing, which will contain sound events characteristic for a large urban agglomeration. The accessibility of public transport for people with disabilities, including people with impaired vision, has for many years been one of the important indicators of mobility of this group of people; not only of those professionally active, but also of those who live independently, including tourists [1–3]. Problems of public transport accessibility for people with disabilities are discussed very widely, and the policy of governments of many countries regarding the mobility of these people deserves special attention [3–5]. The problem of mobility of people with disabilities, including those with impaired vision, requires special preparation of urban space and transport infrastructure, as well as auditory preparation of these people for independent living. To this end, concepts such as special orientation learning methods [6], as well as a hearing-oriented training system for the blind using acoustic virtual reality (VR) based on a head-related transfer function simulation [5] are being created. The use of an intelligent semaphore system (traffic lights) is proposed which communicates with a mobile system carried by the blind person and allows coherent processing of the signals sent and received between the mobile device and the intelligent semaphore [7]. This chapter [8] presents the results of a study on the impact of 3D sound recording and reproduction methods on the efficiency of the location of the sound source by blind and normally sighted people. The efficiency was assessed based on the ability to identify the direction from which the sound comes and the direction of its propagation for the acoustic events of road traffic. Issues related to the anatomy and physiology of hearing, to the imagination skills of the people with impaired vision as well as to devices supporting those people were also investigated. Based on this research, a prototype of an electronic device that supports the orientation of the blind in the environment by means of sound signals was developed [9]. The spectrum of issues related to facilitating the mobility of people with disabilities, especially those visually impaired, is currently very wide. It is impossible to present all publications in this field, therefore only selected ones have been described, showing new trends in research aimed at increasing the mobility of people with impaired vision. In order to develop a novel method of spatial orientation, a targeted grant was obtained, the main purpose of which was to develop a method of teaching spatial orientation in a large urban agglomeration using sounds of the urban environment, designed for people with impaired vision (children and youth). This method was developed in close cooperation with the Special School and Education Centre for Blind Children in Owińska near Poznań. Partners actively declaring to be involved in the project were: Polish Blind Association (District of Wielkopolska), as well as Association for the Blind and Visually Impaired Graduates of the Centre for the Blind 'Być Potrzebnym' in Owińska.

The proposed goal and scope of the project was the first stage of a bigger project dedicated to the development and implementation of the concept of a city that would be safe for people with impaired vision. The scope of the project included, among others:

- surveys concerning the problems of people with impaired vision getting about in urban agglomerations,

- identification and classification of sound sources important in orientation in large urban agglomerations and analysis of the possibility of using vibrations as a signal supporting spatial orientation,
- study on acoustic parameters that will be the basis for the development of new standards for traffic lights and sound signalling at pedestrian crossings,
- elaboration of a method of obtaining and reproducing environmental sounds reflecting real acoustic situations in a city, as well as vibrations—as a supporting signal,
- development of a library of acoustic events for selected places and situations that are necessary for orientation in urban areas,
- a method of teaching people with impaired vision how to spatially orient themselves in a city and verification of this method on the example of the Park of Spatial Orientation in Owińska and the Transport Plan of the City of Poznań.

The end result of the project was a method of teaching people with impaired vision how to spatially orient themselves in a large urban agglomeration, to be used at various levels of the teaching process. In addition, a proposal for guidelines for the standard for traffic lights and sound signalling at pedestrian crossings was developed. As part of the work related to the development of spatial orientation teaching method for people with impaired vision, a survey was conducted to find out the most important problems faced by those people living in a large urban agglomeration and to determine the impact, type and nature of sounds generated in the city in terms of spatial orientation. Next, on the basis of the conclusions resulting from the survey, more than 1700 sound situations characteristic of a large urban agglomeration were selected. The recorded sound signals were subjected to digital analysis with particular emphasis on time-spectral analysis [6, 10, 11].

2 Survey Among People with Impaired Vision

The survey was conducted i.e. with: students of schools for blind people in Poland, blind residents of Poznań (both members of Polish Blind Association and others), blind university students in Poznań, and blind pupils of educational centres for people with impaired vision. The aim of the study was to get to know the most important problems faced by people with impaired vision who live and move around in a large agglomeration. An important issue was also to determine the impact, type and nature of sounds generated in the city on spatial orientation. The scope of analyses and expectations of the survey was focused on the following issues [10]:

- defining problems related to the movement of people with impaired vision around a city,
- defining urban environment that causes the greatest problems with normal movement and orientation of people with impaired vision,
- defining to what extent sound can replace vision and facilitate movement in the environment,

- in which urban situations spatial orientation based on sound is the most desirable and disturbed,
- which sounds of an urban agglomeration pose the greatest problems in spatial orientation.

The conclusions from the conducted surveys were the basis for determining the detailed scope of recording city sounds and for the development of a library of acoustic events. The survey was conducted among 220 blind people. Due to the wide scope and very subjective nature of issues, children and the elderly were asked for opinion.

The survey covered people from the age of 10 to above 61 years. The largest group of people surveyed were people aged between 31 and 40. They constituted 28% of the surveyed people. Another group of respondents were people aged 21–30—20% and very young people and children aged 10–20—18%. The age group from 41–50 years was represented by 11% of the respondents and the group 51–60 years—by 18% of respondents. The age group with the smallest number of people were people aged 61 or more. In this age group, only 5% of respondents were in this age group.

More than half of the people participating in the survey live in cities with more than 100,000 inhabitants. In smaller cities, in subgroups according to the number of inhabitants (from 20,000 to 100,000), the distribution is uniform and equals on average around 10%. Most people who took part in the survey are people who work or study. This answer was declared by 71% of respondents. Only 22% of the respondents declared that they were unemployed. In addition, 7% of people did not work at the time of the survey, but they had already worked in their lives.

In the next question of the survey people were asked how often they leave their home on their own. A summary of the answers to this question is provided in Fig. 1. A very large group of respondents (63%) go out on their own every day. Only 20% of the respondents answered that they leave home a few times a week. There was a group of 11% of the respondents who never leave their home on their own.

The answers to general questions regarding the importance of acoustic information in spatial orientation and movement, interest in learning the ability to interpret the surroundings with hearing and the impact of atmospheric conditions on the perception of sounds of the city can be seen in Fig. 2.

The vast majority of the blind people who participated in the survey believed that changes in the reception of sounds of the urban environment depend on atmospheric conditions (79% of positive answers), as well as that the amount of information about the surroundings obtained through hearing is significant (92%). In addition, 72% of respondents are interested in learning the ability to interpret the environment with use of hearing. These studies confirmed the necessity to create a method of teaching people with impaired vision how to spatially orient themselves using the sounds of the urban environment.

When answering the detailed questions, the surveyed people indicated sounds supporting and hindering spatial orientation. Figure 3 presents a list of sounds that, according to the respondents, may support spatial orientation.

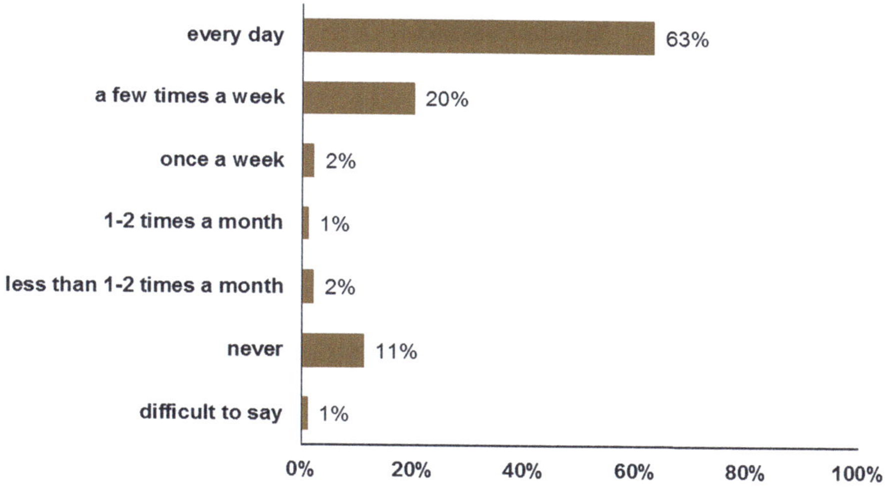

Fig. 1 Frequency of leaving home alone [10]

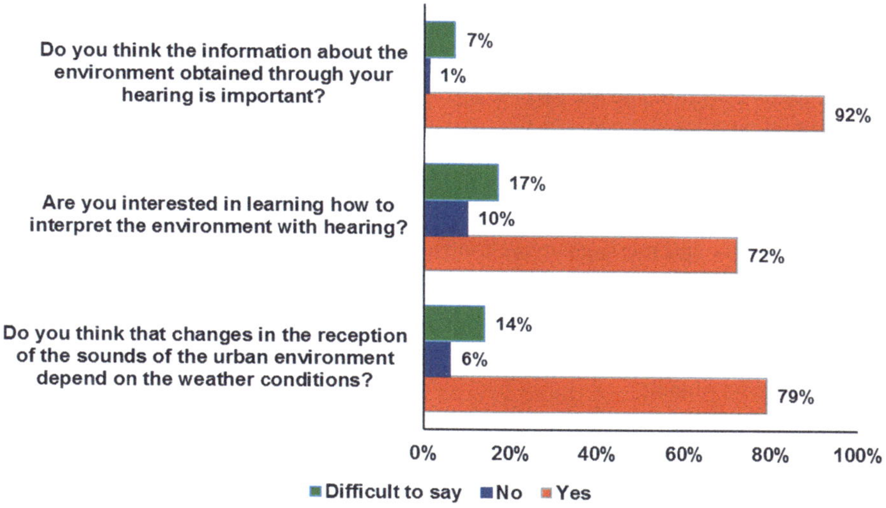

Fig. 2 Significance of acoustic information in the spatial orientation of people with impaired vision [10]

Among the sounds that support spatial orientation, the surveyed people first mentioned pedestrian crossings with sound signals; 69% respondents gave this answer. About 30% was the score of responses related to the means of transport and their infrastructure, i.e. trains, cars and trucks, trams, buses and their stops as well as construction machines. The sounds that support spatial orientation may also include ringing bells in churches, as 26% of respondents said.

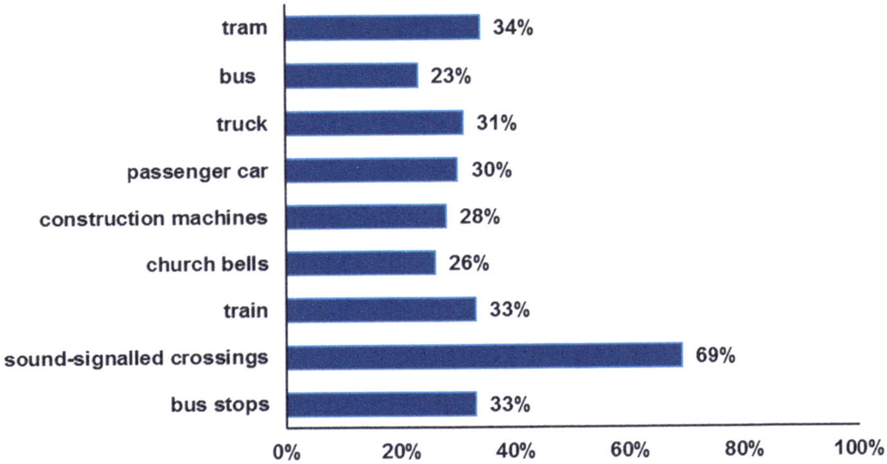

Fig. 3 Sounds supporting spatial orientation [10]

3 Characteristic Places for Acquiring Sounds

Respondents also indicated sounds that impede their spatial orientation in the city and pointed out that it is necessary to register sounds (external and internal) of single passing by of vehicles coming from different directions, standing with the engine running, opening and closing doors, accelerating and braking, sounds in different positions in relation to the stop shelters, sounds generated by behaviour of other passengers and messages from loudspeakers. The respondents named as significant the urban infrastructure sounds related to:

- pedestrian crossings with varying degrees of complexity (with special marking of without it),
- underground tunnels,
- crossing the street outside marked crossing,
- bridges,
- footbridges,
- intersections of various types and sizes on single and multilane streets with and without buildings, with or without a tramway.

From the analysis of questionnaire surveys carried out among people with impaired vision, a number of conclusions were formulated, which were taken into account during recording of acoustic signals in the urban environment. Due to the different age and, as a consequence, the height of people with impaired vision, the acquisition of acoustic signals should take place at two heights according to the height on which the person's ears are located. It was assumed that these were at 0.9 m and 1.60 m from the ground level.

The measurements of sound sources covered also the following means of transport as the main source of noise in the city: passenger cars, buses, trams, passenger and freight trains, motorcycles, heavy vehicles and emergency vehicles (ambulances, fire brigade) as well as situations and events occurring in characteristic places of the city: intersections, roundabouts, etc.

The development of a spatial orientation method for people with impaired vision required obtaining a large number of sound signals (about 1700) that can be found in large urban agglomerations, recorded together with the city background noise. Due to the significant dynamics of acoustic signals occurring in the city and the specificity of the target audience, the recording of acoustic signals was carried out with the use of measuring equipment guaranteeing lossless recording of highly non-stationary phenomena in the full acoustic band. All acoustic signals necessary for the identification and classification of the phenomena for the needs of registration and the method of their application in the 'library—sound bank' were recorded with a single microphone, in real conditions in the city, i.e. with urban background noise.

4 Sound Identification of Selected Objects and Events

In accordance with the adopted assumptions of the target method of presenting recorded signals, the recordings were made simultaneously at two heights: 0.90 m and 1.60 m from the ground level. Measurements at a height of 1.60 m were carried out using a 4100D Head And Torso Simulator from Brüel & Kjær. The second measurement set for recording acoustic signals at a height of 0.90 m in the binaural system were the B&K 4101 in-ear microphones from Brüel & Kjær. Due to the need to teach people with impaired vision to move parallel and perpendicular to the communication routes, it was decided that the recording of sounds will be carried out in two directions: perpendicular and parallel to the sources of the acoustic signal. Figure 4 shows the view of the measuring torso used for the measurements.

After the analysis of the obtained material, objective identification and classification of acoustic phenomena useful in teaching people with impaired vision how to spatially orientate was carried out. The basis for the identification and classification of the phenomena were time-spectral analyses and audition by people with impaired vision—in the role of 'experts' assessing the possibility of identifying individual sources of sounds registered in the urban agglomeration. 'Experts' were people with impaired vision moving around the city on their own. These people represented different groups and different levels of experience according to age, education, place of work and place of residence. Research involving blind experts consisted of reproducing mono audio files, and additionally recorded using in-ear microphones, stereo files with soundtracks, and selected acoustic events of the urban environment.

Due to the need to develop objective criteria for identifying the orientation object, it was decided to perform time-frequency analyses of acoustic signals encountered in the city. Such a decision was made on the basis of conducted comparative research on the possibility of identification of individual objects related to

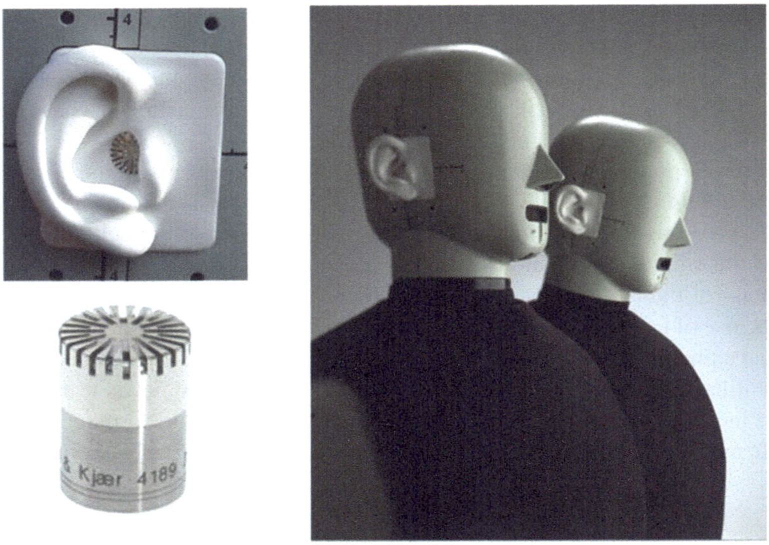

Fig. 4 View of measurement manikin—4100 D HATS Head And Torso Simulator [2]

transport in the city, based on the spectrum of sound generated by these objects [10]. Figure 5a presents an example of the analysis of sound generated by a passenger car, and Fig. 5b shows a passenger car with a trailer. When analysing the time-frequency spectra, it can be seen that each of these vehicles has a different time-spectral structure, and therefore a different signature allowing them to be distinguished. These two examples indicate that it is possible to identify the type and complexity of vehicles (which was confirmed by the experts). The analysis of the spectrum of signals from other vehicles confirmed the possibility of their identification in an urban agglomeration.

Figure 6 shows an example of the analysis of the sound generated by a braking motorcycle, and Fig. 7 by closing doors of a tram.

Figure 8 shows the sound spectrogram (time-spectrum map) generated by a passing ambulance with an acoustic siren on. This is another example of vehicle identification due to the clear audibility in urban traffic and the occurrence of time-frequency modulation, visible in spectrograms in the form of characteristic zigzags parallel to the time axis.

As it results from the sample spectra that identify acoustic events against the background of the city noise, it is possible to fully identify the event and assign the source, which is important for the orientation of the people with impaired vision in the city.

The identification and classification of sound sources (simple and complex) was the basis for the development of a database that was used to: develop a library of acoustic events used in teaching spatial orientation, elaborate the characteristics of

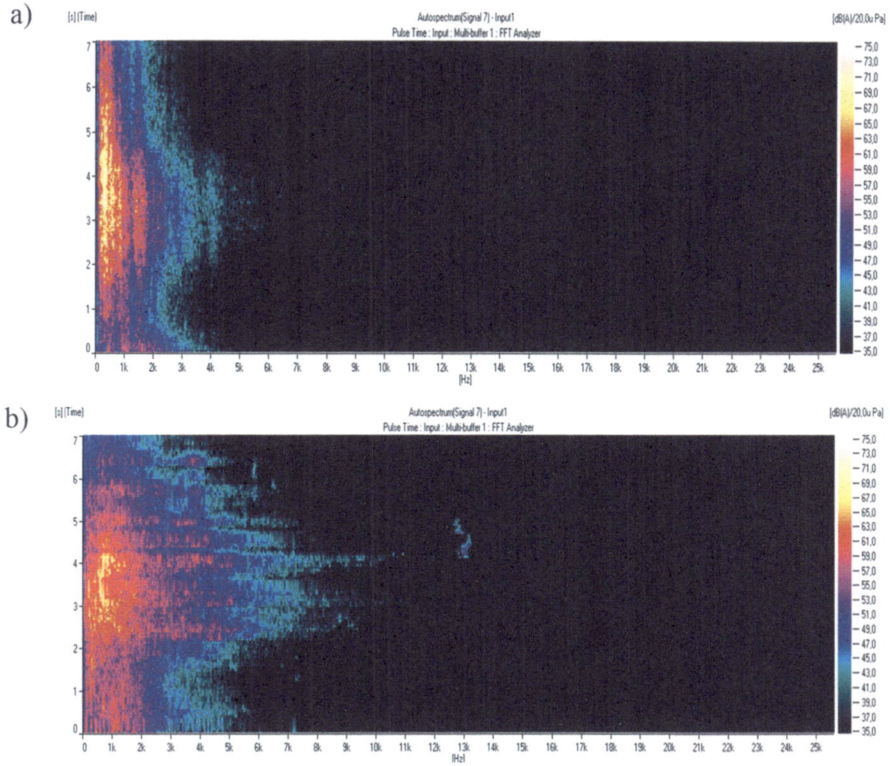

Fig. 5 Time-frequency spectra of a passenger car (**a**) without a trailer, (**b**) with a trailer

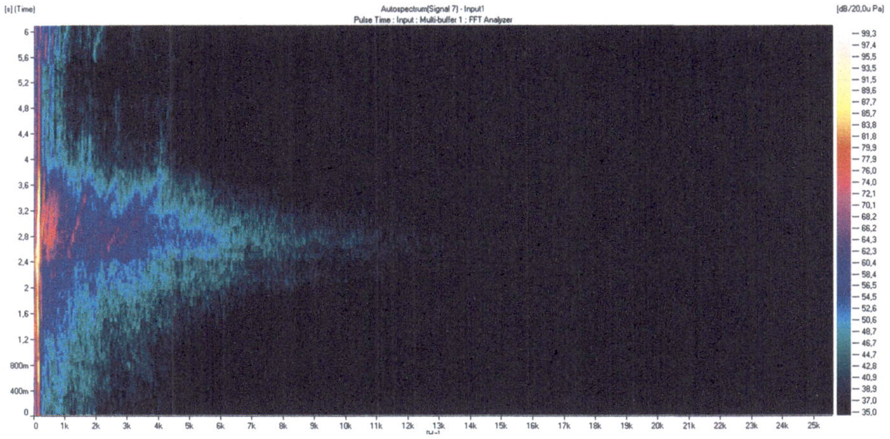

Fig. 6 Time-frequency spectrum recorded during motorcycle braking

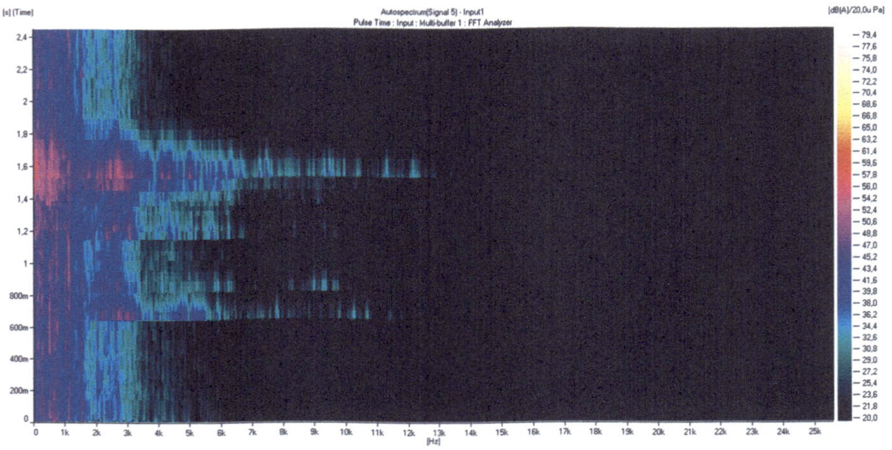

Fig. 7 Time-frequency spectrum recorded during closing tram doors

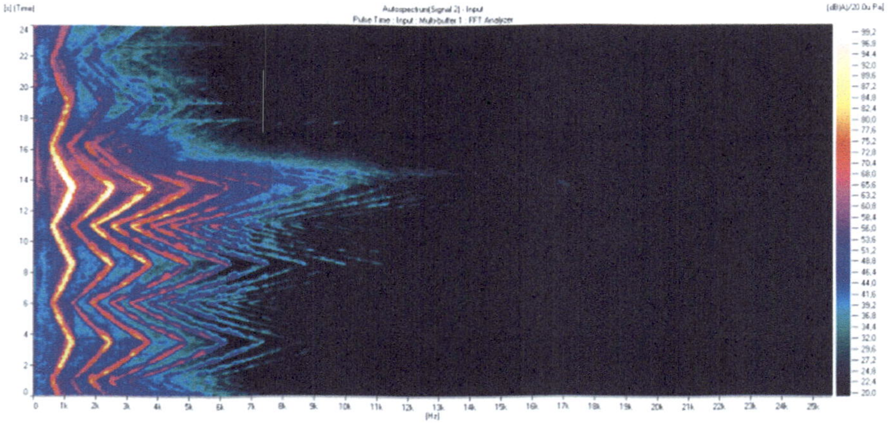

Fig. 8 Time-frequency spectrum of an ambulance passing by

simple and complex sounds and develop assumptions for the spatial orientation teaching method. Thanks to this, people with impaired vision in Poland will get enriched with knowledge and skills that will allow them to move around better and more efficiently in large urban agglomerations, and thus better and more actively operate in the environment of sighted people.

5 Teaching Spatial Orientation to People with Impaired Vision

The aim of the developed teaching method of spatial orientation was that students should acquire skills to move independently in urban environment using the sounds of the city, as well as become motivated to learn and self-educate. This method aims to develop a competence to read urban sounds in a way that enables a walking person to understand the surrounding environment to the maximum and to independently create (perceive) countless routes of safe movement in the city [12]. The general outline of the method is as follows:

5.1 First Level: Evaluation of Simple Events

1. Use of simple recordings for learning: a definition of a street (one and two-lane streets; one and two-way streets), a definition of a road crossing (typical, T-shaped, roundabout, etc.).
2. Determination of the type of vehicles moving in urban environment (cars, trams, buses and emergency vehicles).

5.2 Second Level: Evaluation of Complex Events

1. Getting to know intersections (of all types), with traffic lights (including sound signalling) and without traffic lights.
2. Finding the location of public transport stops—using reflections of sounds (shelter), underground passages (roofed), kiosks at tram stops.
3. Use of complex sounds, among others, to: search for pedestrian crossings, determine the types and widths of streets, determine the location of tram tracks, location and speed of vehicles moving with constant speed, braking, accelerating, turning, etc.

5.3 Third Level: Evaluation and Association of Complex Events

1. Use of recordings of streets characterised by various widths, configurations and traffic intensity, as well as the description of large intersections with the use of sound recording.
2. Interpretation of communication situations, infrastructure layout and layout of the environment in case of disturbances.
3. Interpreting traffic in difficult weather conditions (wind, rain, slush, etc.).

6 Conclusions

This chapter presents issues related to the selection and analysis of sound signals that are important in the development of the method of spatial orientation in a large city, designated for people with impaired vision. The need to develop such a method resulted from the fact that a large part of visually impaired people actively participate in social life and hearing may be a substitute for their eyesight. Identification and classification of urban agglomeration signals made on the basis of an objective analysis of the physical features of the recorded signals confirmed the possibility of distinguishing individual situations and acoustic events indicated by the blind in the survey.

Acknowledgements Presented research and the paper are partly funded by Statutory Activities fund of the Institute of Combustion Engines and Transport, PUT (PL) 5/52/DSPB/1277.

References

1. Marciszewska, E., Bergel, I., Matuška, J., Záhorová, Y.: Przegląd Komunikacyjny, No. 3/2018, pp. 14–21
2. Nicklas, J.P.G., Schluter, N., Schnieder, L., Winzer, P.: Entwicklung einer anforderungsgerechten Mobilitatsassistenz fur in ihrer Mobilitat eingeschrankte Reisende. Oualitatsmethoden im Diskurs zwischen Wissenschaft und Praxis. **18**, 229–243 (2015)
3. Bergel, I.: Handicapped Persons in the Transport Policy of the Polish State, Conference Proceedings, vol. 1. TRANSED 2001, Warsaw, Poland, 2–5 July 2001 (2001)
4. Butlewski, M., Kalemba, A., Sydor, M.: Wymagania dla miejskich systemów transporowych wobec rozwoju sprzętu dla osób z niepełnosprawnościami [Requirements for urban transport systems towards the development of equipment for people with disabilities]. Logistyka 6/2014
5. Seki, Y., Sato, T.: A training system of orientation and mobility for blind people using acoustic virtual reality. IEEE Trans. Neural Syst. Rehab. Eng. **19**(1) (February 2011)
6. Opracowanie metody orientacji przestrzennej w dużym mieście dla osób niewidomych z wykorzystaniem dźwięków środowiska [Development of a spatial orientation method in a large city for blind people using environmental sounds]: Final report. Development grant in cooperation with the Adam Mickiewicz University in Poznań, project no. 0393/R/H03/2008/04
7. Montanha, A., Escalon, M.J., Dominguez-Mayo, F.J., Polidorio, A.M.: A technological innovation to safely aid in the spatial orientation of blind people in a complex urban environment. International Conference on Image, Vision and Computing (2016)
8. Furmann, A., Skrodzka, E., Giżewski, Ł., Nowotny, P.: The effect of sound reproduction method on performance in sound source localization by visually impaired and normally sighted subjects. Acta Phys. Polonica **123**(6) (2013)
9. Kubanek, M., Bobulski, J.: Device for acoustic support of orientation in the surroundings for blind people. Sensors. **18**, 4309 (2018). https://doi.org/10.3390/s18124309
10. Orczyk, M.: Assessment of internal and external noise of the urban transport system. Wydawnictwo Politechniki Poznańskiej [Publishing House of the Poznan University of Technology] 2019. ISBN 978-83-7775-534-1

11. Orczyk, M., Tomaszewski, F.: Badania i analiza widmowo-czasowa hałasu na skrzyżowaniu typu rondo [Measurement and spectral-time analysis of noise at a roundabout]. In: XLIV Zimowa Szkoła Akustyki Środowiska i Wibroakustyki [XLIV Winter School of Acoustics of the Environment and Vibroacoustics], Gliwice-Szczyrk (2016)
12. Czechyra, B., Tomaszewski, F., Skrodzka, E., Durek-Sypek, I.: New method of spatial orientation teaching based on city sounds, not only for the blind. In: The 17th International Congress on Sound and Vibration, Vilnius, Lithuania, 8–12 July 2012.

Index

M. Suchanek (ed.), *Transport Development Challenges in the 21st Century*, Springer Proceedings in Business and Economics,
https://doi.org/10.1007/978-3-030-50010-8

The manufacturer's authorised representative in the EU is Springer Nature Customer Service Centre GmbH, Europaplatz 3, 69115 Heidelberg, Germany. If you have any concerns regarding our products, please contact ProductSafety@springernature.com

Printed and bound by CPI Group (UK) Ltd, Croydon, CR0 4YY
30/09/2026
02225692-0007